Protoplasmatologia

Begründet von / Founded by

L. V. Heilbrunn, Philadelphia, Pa. · F. Weber, Graz

Herausgegeben von / Edited by

M. Alfert, Berkeley, Calif.
H. Bauer, Tübingen

W. Sandritter, Freiburg i. Br.
P. Sitte, Freiburg i. Br.

Mitherausgeber / Advisory Board

J. Brachet, Bruxelles
D. Branton, Berkeley, Calif.
H. G. Callan, St. Andrews
W. W. Franke, Freiburg i. Br.

N. Kamiya, Osaka
G. F. Springer, Evanston, Ill.
L. Stockinger, Wien
B. F. Trump, Baltimore, Md.

II Cytoplasm
E Cytoplasm Surface
1 Membrane Structure

Springer-Verlag
Wien New York 1972

Membrane Structure

D. Branton and D. W. Deamer

With 24 Figures

Springer-Verlag
Wien New York 1972

This work is subject to copyright.
All rights are reserved, whether the whole or part of the material is concerned, specifically those of translation, reprinting, re-use of illustrations, broadcasting, reproduction by photocopying machine or similar means, and storage in data banks.
© 1972 Springer-Verlag/Wien
Library of Congress Catalog Card Number 70-170378
Printed in Austria by Adolf Holzhausens Nfg., Wien

ISBN 3-211-81031-5 Springer-Verlag Wien - New York
ISBN 0-387-81031-5 Springer-Verlag New York - Wien

Vor 20 Jahren haben Professor FRIEDL WEBER, Graz, und Professor L. V. HEILBRUNN, Philadelphia, das Handbuch „Protoplasmatologia" gegründet, zu einer Zeit, als es der Stand der Wissenschaft noch erlaubte, an ein Handbuch im klassischen Sinne zu denken, das heißt an eine umfassende Darstellung des Gesamtgebietes. Die ursprüngliche Disposition sah zwölf Bände mit einer hierarchischen Untergliederung vor. Die rasche Entwicklung auf diesem Gebiet mit der zunehmenden Differenzierung einerseits und der Bildung neuer Schwerpunkte andererseits hat im Lauf der Zeit mehrfache Änderungen der Disposition notwendig gemacht. Neue Gesichtspunkte ergaben sich auch durch den Wechsel im Herausgeberkollegium nach dem Tod der Gründer.

Seit dem Jahre 1953 sind 51 Einzelbände mit einem Gesamtumfang von rund 9400 Seiten erschienen.

Die Herausgeber haben nun im Einvernehmen mit dem Verlag beschlossen, die engen Schranken der früheren Handbuchdisposition zu verlassen. In einer zeitgemäßeren Form sollen Ergebnisse und Probleme der Zellbiologie in Monographien dargestellt werden. So wird es möglich sein, jeweils besonders aktuelle Themen zu behandeln. Erhalten bleiben wird der Anspruch auf höchstes wissenschaftliches Niveau.

Herausgeber und Verlag

Twenty years ago, Prof. FRIEDL WEBER (Graz University) and Prof. L. V. HEILBRUNN (University of Pennsylvania) conceived the idea for the handbook "Protoplasmatologia" at a time when the state of knowledge in the field of cell biology still permitted one to think of an all-encompassing handbook in the classical sense. Since 1953 fifty-one volumes with a total of about 9,400 pages have been published. The very rapid developments in this area of science, especially during the last decade, have led to new insights which necessitated some alterations in the original plan of the handbook; also, changes in the board of editors since the death of the founders have brought about a reorientation of viewpoints.

The editors, in agreement with the publisher, have now decided to abandon the confining limits of the original disposition of the handbook altogether and to continue this work, in a form more appropriate to current needs, as an open series of monographs dealing with present-day problems and findings in cell biology. This will make it possible to treat the most modern and interesting aspects of the field as they arise in the course of contemporary research. The highest scientific, editorial and publishing standards will continue to be maintained.

Editors and publisher

Protoplasmatologia
II. Cytoplasma
E. Cytoplasma-Oberfläche
1. Membrane Structure

Membrane Structure

By

DANIEL BRANTON

Department of Botany, University of California, Berkeley, CA 94720, U.S.A.

and

DAVID W. DEAMER

Department of Zoology, University of California, Davis, CA 95616, U.S.A.

With 24 Figures

Contents

	Page
I. Introduction	2
II. General Organization of the Membrane Matrix	3
A. Physical Restrictions on Membrane Structure	3
1. Total Membrane Area and Volume	4
2. Area Occupied by Lipid	4
3. Membrane Surface Tension	5
B. Membrane Models	6
1. Protein-Lipid-Protein Membranes	7
a) Myelin Birefringence	8
b) X-Ray Diffraction of Myelin	8
c) X-Ray Diffraction of Other Membranes	10
d) Electron Microscopy of Membrane Thin Sections	12
e) Lipid and Protein Measurements	14
f) Thermal Analysis	18
g) Electron Spin Resonance	20
h) Criticism of the Protein-Lipid-Protein Model	22
2. Lipid-Protein-Lipid Membranes	22
a) Infrared Spectroscopy	23
b) Optical Rotary Dispersion and Circular Dichroism	25
c) Nuclear Magnetic Resonance	26
d) Enzymatic Hydrolysis of Membrane Components	28
e) Criticism of the Lipid-Protein-Lipid Model	30
3. Particulate Membranes	30
a) Chloroplast Membranes	31
b) Other Membranes	33
c) Disassembly-Reconstitution Experiments	35
d) Criticism of the Particulate Membrane Hypothesis	36
4. Conclusions about Membrane Models	37
III. Differentiation and Specialization of Membranes	38
A. Differentiation of Boundary and Enzyme Functions	39
B. Freeze-Etching	41
IV. Conclusions	59
V. Bibliography	61

I. Introduction

The boundary of all cells is delineated by a thin membrane which separates the protoplasm from its environment. The existence of this membrane was first suggested by DE VRIES (1885), PFEFFER (1897) and OVERTON (1899), who demonstrated that plant cells respond osmotically to variations in external osmolarity. However, the physical reality of the cell membrane remained controversial for many years. Cell membranes are too thin to be directly visualized in the light microscope, and their boundaries were confused with other structures at the cell surface such as the cytoplasmic cortex in animal cells or the cell wall in plant cells. Therefore, early notions about membrane structure were indirectly derived from studies of membrane function, and in fact the chemical composition of the plasma membrane was inferred from its permeability properties before any direct analyses of individual membranes could be attempted.

Around 1930, new approaches became available for the study of cellular membranes. These included micromanipulation, surface chemistry techniques, electrical measurements, X-ray diffraction and electron microscopy. The existence of cell membranes was soon directly and unambiguously demonstrated first by micromanipulation (PLOWE 1931) and later by a number of investigators who used positive staining techniques of electron microscopy (for review see ELBERS 1964). Electron microscopy further revealed the astonishing fact that most intracytoplasmic components, including chloroplasts, mitochondria, nuclei, Golgi bodies, vacuoles and endoplasmic reticulum, are also bounded by a membrane which seemed, at least superficially, to resemble the cell membrane. Most membranes were found to be 60 to 100 Å thick and usually appeared as two dark lines separated by a lighter space ("railroad tracks") when sectioned normal to their plane (ROBERTSON 1959, WHALEY et al. 1960).

With the growing knowledge of cell ultrastructure came techniques for isolating the intracytoplasmic organelles. Combined biochemical and electron microscopic investigations soon established that many cellular activities were compartmented within or mediated by membrane bound organelles. It is now common knowledge that active transport, selective permeability, quantum conversion in photosynthesis, oxidative phosphorylation and nerve conductivity are mediated by specific biological membranes (ROODYN 1967, DE THE 1968, LESLIE 1968, SINGER and TASAKI 1968).

When relatively pure preparations of isolated organelles became available, the composition of many membranes was established. A unique and differently proportioned set of lipids and proteins was found to be characteristic of each membrane type (BENSON 1964, KORN 1966, O'BRIEN 1967). However, the manner in which this compositional specificity forms the basis for the functional specificity of membranes has yet to be elucidated.

During the past few years a number of refined spectroscopic and calorimetric techniques have been used to analyze both the organization and the motion of molecules within biological membranes. These physical techniques provide data on the conformation of membrane protein and lipid *in situ*, and have been valuable in establishing the bulk structure of membranes. However in spite of our growing knowledge of membrane function and composition, the physiological capabilities of a membrane cannot be explained in terms of its molecular structure. We know the functions of many membranes and we have some understanding of their structure and composition, but we don't understand enough about the molecular ordering of their biochemical components to account for their specific activities.

There are certainly a number of reasons for this ignorance. In the first place, detailed studies of membrane composition have only recently been possible because pure preparations of a single membrane type were not previously available. Secondly, membrane protein tends to be insoluble and therefore difficult to characterize. Thirdly, membrane phenomena occur in or on a highly structured matrix and therefore are not amenable to standard techniques of solution chemistry. Although physical techniques using X-ray diffraction, optical rotation or paramagnetic resonance have proven to be powerful tools for studying highly structured systems such as membranes, these methods are essentially averaging techniques. Functionally important centers within a membrane may be too few or too transient to provide the structural organization or periodicity detectable by averaging techniques. Thus, these procedures may provide a clear picture of the overall membrane matrix without indicating much about the structure and distribution of its most important functional elements.

Many students of membrane biology have assumed that the structural parameters which contribute to the general organization of membrane components must also explain membrane function. As we shall show, calculations based on available evidence do not support this assumption and suggest instead that only a small fraction of the membrane's area may account for most of its specialized functions. Therefore, in the first sections of this review we consider the overall organization of the membrane matrix apart from any elements which may account for its specialized functions. In the second part of the review we consider evidence for differentiation within the membrane matrix, and we show how differentiated or specialized elements in the plane of the membrane are characteristic of specific membrane types. Other aspects of membrane structure have recently been reviewed (ROBERTSON 1964, KAVANAU 1965, VAN DEENEN 1965, FINEAN 1966, MADDY 1966, STEIN 1967, GLAUERT and LUCY 1968, ROTHFIELD and FINKELSTEIN 1968, SJOSTRAND 1968, KORN 1969, STOECKENIUS and ENGELMAN 1969, THOMPSON and HENRY 1970).

II. General Organization of the Membrane Matrix

A. Physical Restrictions on Membrane Structure

Although the problem of resolving biological membrane structure at the molecular level may at first seem hopelessly complex, there are a number of simple physical parameters which restrict the arrangement of lipid and protein molecules within a membrane. These restrictions allow us to reduce the number of possible general models and to apply experimental tests to choose between the remaining alternatives. These restrictions, although obvious, have not been stressed in the past, probably because qualitative and quantitative knowledge of membrane composition was rudimentary. In view of the fairly precise estimates of lipid and protein contents which are currently available for several membranes, we may now consider three major restrictions which help define alternative membrane structures:

1. Total Membrane Area and Volume

It is obvious that membrane thickness (T), surface area (A), volume (V), mass (M) and density (ρ) are related by several simple equations:

$$V = A \times T \tag{1}$$
$$\rho = M/V \tag{2}$$
$$A = M/\rho T \tag{3}$$

These relations may be tested in the erythrocyte membrane where all the parameters have been measured with a fair degree of accuracy. For instance, calculations based on measurements of human erythrocyte diameters show that the surface area is approximately $145 \pm 8 \, \mu^2$ (Ponder 1948, Westerman et al. 1961). A comparable value should be derived from Equation (3) given values of M, ρ, and T. Dodge et al. (1963) found 5.2×10^{-10} mg lipid and about 6.0×10^{-10} mg protein per erythrocyte ghost. About 15 per cent of their hydrated mass is water (Finean et al. 1966). Electron microscopic evidence suggests that the $KMnO_4$-fixed erythrocyte membrane is 75 Å thick. Correcting for a 10 per cent shrinkage which occurs during fixation and preparation for electron microscopy (Fernandez-Moran and Finean 1957, Moretz et al. 1969), the hydrated membrane is about 80 Å thick. The density of hydrated erythrocyte membranes is about $1.13 \, \text{g cm}^{-3}$ (Steck et al. 1970). Substituting in Equation (3):

$$A = \frac{11.2 \times 10^{-13} \text{g lipoprotein} + 1.7 \times 10^{-13} \text{g water}}{1.13 \, \text{g cm}^{-3} \times 80 \times 10^{-8} \, \text{cm}} =$$
$$143 \times 10^{-8} \, \text{cm}^2 = 143 \, \mu^2.$$

This figure compares favorably with the $145 \pm 8 \, \mu^2$ estimate of surface area and suggests that calculations of this type may be used to estimate the surface area of membranes when their thickness, lipid and protein content, and density are known.

2. Area Occupied by Lipid

Membrane lipid content is relatively easy to measure, since lipids are readily extracted and qualitative and quantitative determinations by chromatography are now routine. Obviously lipids must occupy a certain area and volume within a membrane. If the lipids are extracted from a membrane preparation and spread as a monolayer, it is possible to estimate the area which might be occupied by the lipid within the membrane (Gorter and Grendel 1925, Bar et al. 1966). To be sure, such data can only provide a rough estimate. The smallest area occupied by a molecule of extracted lipid in a monolayer is the area at collapse pressure. This is more or less equivalent to the area occupied by the molecule in a crystal. The largest area can be taken as the area occupied by the film when the first detectable surface pressure is measured. Thus the area occupied by a lipid molecule in a monolayer is not fixed but depends on how much the film is compressed. Nevertheless, it is possible to measure the areas of monolayers of extracted lipids and determine whether the measurements are consistent with a postulated number of lipid layers in the original membrane. For instance, one may ask whether the area occupied by a monolayer either at collapse pressure or at the lowest detectable surface pressure is sufficient to form a bilayer of lipid in the membrane from which it was extracted.

The general equations involved in these calculations are shown below:

$$\text{Number of lipid layers} = \frac{\text{Area of extracted lipid (monolayer)}}{\text{Area of membrane}} \quad (4)$$

$$\frac{\text{Area per lipid molecule}}{\text{assuming a bilayer}} = \frac{\text{Area of membrane} \times 2}{\text{Number of molecules}} \quad (5)$$

These equations will be used later to calculate the area occupied by lipid in several types of membranes.

3. Membrane Surface Tension

Various methods have been devised for measuring the interfacial tension at membrane surfaces (for a review see HARVEY 1954). The results always show values at least one order of magnitude below those for a water-hydrocarbon interface. For example, HARVEY and SHAPIRO (1934), using a microscope centrifuge, photographed the flattening of oil containing vesicles in the cytoplasm of a mackerel egg against the cell membrane. Using the sessile drop equation they calculated the tension at the oil droplet-cytoplasm interface to be 0.6 dynes/cm. The surface tension of mackerel body oil against several aqueous solutions was measured with a du Nouy tensimeter. The lowest value obtained was 10.0 dynes/cm. The only way to explain the low interfacial tension of the oil droplet is to assume that it is surrounded by a membrane with a hydrophilic surface. Similar measurements show that all membranes exhibit a low interfacial tension against water. These low values would only be found if the membrane surface were hydrophilic. The surface tension measurements therefore place a third restriction on membrane structure, namely that most of the hydrophobic membrane components—hydrocarbon chains and apolar amino acid side chains—must be directed inwardly, leaving hydrophilic groups at the membrane surface to interact with the surrounding aqueous phase. However, this restriction does not specify whether it is the hydrophilic groups of the proteins or of the lipids which are at the membrane surface.

B. Membrane Models

With these restrictions in mind we may examine the two basic models which have been proposed for the molecular organization of membranes. The first is based on a continuous lipid bilayer with protein at the membrane surface on both sides of the bilayer (Fig. 1 a). We shall term this a protein-lipid-protein (PLP) model. According to the PLP model, the hydrophilic groups of the proteins are at the surface of the membrane. The second model proposes either a lipid-protein mosaic or a protein lattice interpenetrated by lipid. According to this model many of the lipid hydrophilic groups are at the membrane surface in direct contact with the surrounding aqueous phase (Fig. 1 b). We shall term this a lipid-protein-lipid (LPL) model. Variants of the LPL model propose that the lipid-protein structure is composed of distinct subunits which aggregate in two dimensional arrays to form the membrane fabric (Fig. 1 c).

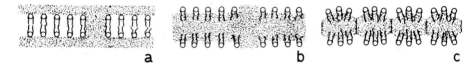

Fig. 1. Membrane models. Generalizing some of the more detailed structures which have been proposed, these diagrams stress the relative arrangement of lipid and protein: a) protein-lipid-protein (PLP), b) lipid-protein-lipid (LPL) and c) subunit schemes for the biological membrane. From BRANTON (1969).

Certainly the diagrams of Fig. 1 are gross oversimplifications of some of the elegant membrane models which have been proposed, but even if we neglect these diagrammatic simplifications, it appears improbable that attempts to assign each molecule a unique place in the membrane matrix can succeed. The structure of the membrane, and particularly the relation of the proteins to each other or to the membrane lipids, is not static. A recent paper by FRYE and EDIDIN (1970) provides a striking example of the dynamic fluidity which must be involved. Cell lines of mouse and human origin were fused and the hybrids treated with fluorescent antibodies (red for human and green for mouse antigens). Shortly after fusion, most hybrids showed distinct red and green halves, but within an hour the colors were completely mixed. The results indicated diffusion and relative change of position of large molecular components of the cell membranes and showed that structure within a single membrane certainly varies with time. We must therefore recognize that any membrane model is at best a time-averaged representation of molecular architecture.

The time-averaged molecular organization of most membranes probably falls within the spectrum of the general models to be discussed. Since all of these models can be made to fit the restrictions on total area and volume, lipid area, and surface tension discussed above, we shall examine specific membranes to determine whether any one of these models more closely reflects the properties of biological membranes than do the others.

1. Protein-Lipid-Protein Membranes

In 1935 DANIELLI and DAVSON suggested a molecular model for the cell membrane consisting of a very thin lipid layer with protein adsorbed upon

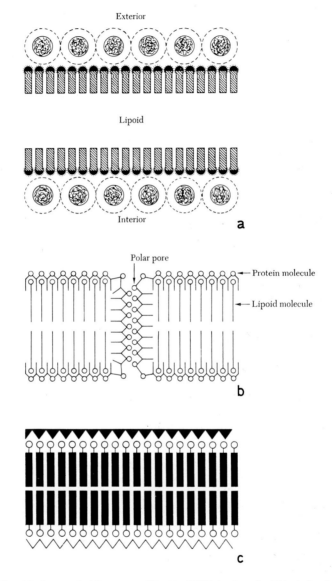

Fig. 2. *PLP* models. *a*) The original proposal of DANIELLI and DAVSON (1935) showing a lipoid sheet of undefined thickness. *b*) DANIELLI'S modification (1954) showing a lipid bilayer and hydrophylic pores. *c*) Diagram of the unit membrane (ROBERTSON 1959) showing the relation of this concept to other *PLP* models.

both surfaces (Fig. 2 a). They argued that such a membrane would, in theory, be capable of distinguishing between molecules of different size and solubility characteristics and could have other functional attributes known to exist at the cell surface. This protein-lipid-protein (PLP) model was subsequently refined to indicate a membrane of defined thickness based on a continuous central lipid bilayer (DANIELLI 1954) (Fig. 2b). This model assumes that the interaction between the lipid and protein is dominated by electrostatic forces which would cause the protein to unroll into a β-pleated sheet configuration at the polar surfaces of the lipid bilayer. This paucimolecular model of DANIELLI has profoundly influenced the thinking of biologists and has engendered numerous important biochemical and biophysical studies. The examination of myelin membranes has been particularly important as myelin provides a unique, naturally ordered system of membranes which can be examined by a variety of radiant probing techniques, including optical birefringence, X-ray diffraction and electron microscopy. Because so many of our current concepts of membrane structure were developed by correlating these analyses of myelin membranes, it is worth reviewing the history of this research in some detail.

a) Myelin Birefringence

Interest in the anatomy and physiology of the nervous system led to extensive light and polarization microscope studies of myelin over a quarter of a century before it became known that this tissue was composed of membranes. Polarization microscopy demonstrated that nerve tissue was strongly birefringent (WYNNE 1900), and the high degree of molecular organization indicated by these findings attracted considerable attention. However, it remained for SCHMIDT (1936) to derive a reasonable model of the myelin sheath which accounted for both its strong, positive, intrinsic birefringence and its weak, negative, form birefringence. SCHMIDT postulated that layers of lipid molecules oriented with their long axes perpendicular to the surface of the myelin sheath alternated with protein lamellae layered parallel to the surface of the sheath. Although some aspects of SCHMIDT's model have proven incorrect, the general molecular orientation that he attributed to the lipids and proteins has been supported by X-ray diffraction and is incorporated in many current notions of membrane architecture.

b) X-Ray Diffraction of Myelin

SCHMITT *et al.* (1935) were the first to examine the low-angle X-ray diffraction pattern of myelin. They discovered an unexpectedly large spacing, ca. 170 Å in frog myelin, that suggested the dimensions of the basic repeating structures postulated by SCHMIDT. Subsequent diffraction data and further analysis led SCHMITT *et al.* (1941) to conclude that the myelin sheath is composed of lipid bilayers wrapped concentrically about the nerve axon with protein intercalated between the lipid bilayers. Thus, SCHMITT *et al.*'s model based on X-ray diffraction data provided powerful confirmation of SCHMIDT's model derived from birefringence studies.

Cogent as were these proposals concerning the molecular architecture and repeat dimensions within the myelin sheath, no conclusions relating these dimensions to those of the biological membrane could be made until GEREN (1954) and later ROBERTSON (1959) provided convincing electron microscope evidence that the layered myelin sheath membranes were directly derived from and continuous with the Schwann cell membrane as diagramed in Fig. 3 a.

GEREN's suggestion that myelin membranes and cell membranes may be similar in molecular composition and arrangement led FERNANDEZ-MORAN and FINEAN (1957) to correlate the electron microscope image of the myelin sheath with its low-angle X-ray diffraction pattern. This study was among

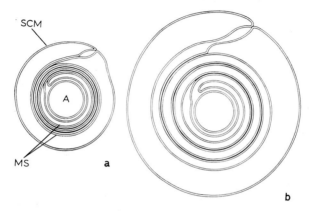

Fig. 3. The myelin sheath (*MS*) develops around the axon (*A*) in continuity with the Schwann cell membrane (*SCM*). a) The major repeat represents the apposition of cytoplasmic surfaces, the intraperiod repeat represents the apposition of outer surfaces. b) Swelling in hypotonic medium separates these outer surfaces but not the cytoplasmic surfaces. From ROBERTSON (1959 and 1964).

the first to regard myelin explicitly as a model system for the study of cell membranes, and it established the framework for more detailed X-ray diffraction and electron microscope analyses.

While FERNANDEZ-MORAN and FINEAN were primarily concerned with the spacings revealed by the positions of their X-ray diffraction maxima, FINEAN (1962) and FINEAN and BURGE (1963) related the magnitudes or intensities of the diffraction maxima to the electron density and hence to the position of atoms and molecules within the repeating myelin membrane (Fig. 4 a). Although their results presented only the most reasonable of several mathematically viable solutions, they have received convincing support from X-ray diffraction studies of myelin swollen in hypotonic solution. This swelling separates the outer surfaces of the two adhering plasma membranes (ROBERTSON 1964) (Fig. 3 b) and is the basis for an X-ray analysis which could lead to a unique transform, and hence a unique structural interpretation, of nerve myelin (WORTHINGTON and BLAUROCK 1969 a) if the molecular structure of swollen and intact myelin membranes were identical. The swelling techniques have been used to obtain parameters for a model based on diffraction data to about 30 Å spacing. The model shows two non-symmetric triple layered membrane units, each trilaminar unit representing a membrane whose electron densitiy distribution is similar to that expected from a PLP model (FINEAN and BURGE 1963, WORTHINGTON and BLAUROCK 1969 b).

More detailed and direct determinations of electron density profiles in three types of myelin have recently been obtained by CASPER and KIRSCHNER (1971). Their determination was based on the condition that a similar structure accounts for the diffraction patterns of both sciatic and optic myelins. By comparing intensity measurements on their various diffraction patterns, which extended to 10 Å spacing, they were able to derive a unique electron density profile of such detail that it could be used to define the location of distinct chemical components of the membrane (Fig. 4 b). Previous work had indicated the general bilayer arrangement of membrane lipids (Fig. 4 a). Extending and confirming these ideas, CASPAR and KIRSCHNER (1971) concluded that the methyl ends of the hydrocarbon chains are distributed within a central layer about 15 Å thick. The low electron density in this central zone rules out the possibility of significant amounts of protein in this region of the membrane although small amounts extending across the bilayer could not be ruled out. The hydrocarbon thickness between the fatty acid polar groups was judged to be about 38 Å in rabbit myelin and 35 Å in frog myelin. The density measurements indicated that the portions of the hydrocarbon chains closest to the polar groups are predominantly close packed, whereas the ends near the bilayer center are pliant and disordered. The protein and water are distributed in the spaces between the lipid bilayers. A most striking feature of their high resolution profiles is the asymmetry of the steps on either side of the central hydrocarbon bilayer region (Fig. 4 b). CASPAR and KIRSCHNER identify these steps with the steroid portion of the myelin cholesterol. They suggest that cholesterol is concentrated in the external (outer) layer of the hydrocarbon bilayer where it is estimated that there is an equimolar ratio of cholesterol and polar lipids, They estimate a ratio of about 3 : 7, cholesterol : polar lipids, in the inner, cytoplasmic layer.

c) X-Ray Diffraction of Other Membranes

Although other membrane systems have not been explored in nearly the detail possible in myelin, X-ray diffraction images have recently been obtained from red blood cell ghosts, microsomal membranes, and isolated mitochondrial membranes packed and oriented by strong centrifugal forces and controlled dehydration (COLEMAN and FINEAN 1966, THOMPSON et al. 1967, FINEAN et al. 1966, FINEAN et al. 1968). The diffraction patterns obtained during controlled dehydration suggest that a certain amount of water can be removed from between and around the membranes without affecting their structure. Both the low-angle and the wide-angle diffraction patterns from these artifically oriented membrane systems are similar to those of the naturally oriented myelin membranes. Assuming that the dehydration techniques used in these studies do not alter membrane structure, the similarities of the X-ray reflections from a wide variety of membranes indicate the essential structural similarities of these membranes and suggest that many

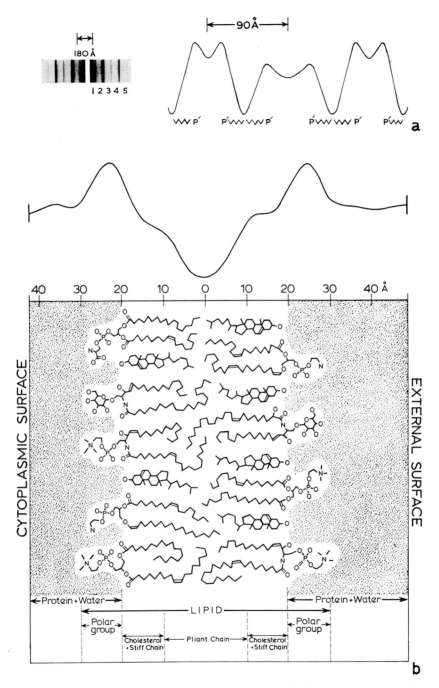

Fig. 4. a) X-ray diagram (left) and calculated electron density profile (right) of rat sciatic nerve myelin. The regions of high electron density are associated with the phosphate head groups and proteins, whereas the regions of low electron density are associated with the —CH$_2$-groups of the membrane lipids. From FINEAN (1962). b) A more detailed electron density profile (top) and schematic (bottom) of rabbit sciatic myelin. From CASPAR and KIRSCHNER (1971).

membranes consist of a lipid bilayer with non-lipid components associated with both surfaces. X-ray diffraction patterns of chloroplast lamellae, however, may differ from those of other membranes and will be discussed in greater detail below.

Thus far we have been discussing X-ray studies of parallel, regularly spaced stacks of membranes. These arrays concentrate the diffracted X-rays in sharp, easily recorded reflections which can be used to determine the spacings and relative orientations of different parts of the membrane. However, these sharp reflections are often difficult to interpret unless their informational content is supplemented by special preparatory procedures, such as swelling. Furthermore some membranes cannot be stacked, and, in those that can, the drying techniques used may alter membrane structure (Finean et al. 1966 and 1968). For these reasons, Wilkins et al. (1971) have developed X-ray analysis methods which can be used on random dispersions of membrane fragments.

Wilkins et al. (1971) have applied these methods to sonicated aqueous dispersions of phospholipids as well as to suspensions of *Mycoplasma laidlawii* membranes, erythrocyte ghosts, nerve-ending plasma membranes, isolated sarcoplasmic reticulum and *E. coli* plasma membrane vesicles. The phospholipid dispersions examined were lipid bilayers in the form of spherical, water-filled vesicles. They provided a model system with recognizable X-ray diffraction patterns that helped to interpret the membrane patterns. The diffraction from all of the biological membranes produced transforms closely resembling those of the lipid bilayers. *Mycoplasma laidlawii* membranes provided the clearest evidence for the existence of a bilayer of lipid. In particular, the thickness of the bilayer, measured from the positions of the diffraction bands, agreed with that calculated for the lipids known to be present in the membrane. Furthermore, the thickness, as indicated by the band positions, increased in proportion to the average length of the hydrocarbon chains incorporated into the membranes as would be expected for a bilayered structure. In all membranes examined the main diffraction bands and the submultiple bands and thickness values derived from the band positions were in rough agreement with a bilayer model. Wilkins et al. point out, however, that their results do not necessarily imply that the bilayer is continuous throughout the membrane. In fact, using their measurements for the distance between lipid phosphate groups across an erythrocyte membrane (45 Å) and the average area per lipid molecule (117 Å^2, Engleman 1969 a), they estimated that between 25 per cent and 30 per cent of the lipid bilayer surface area may be occupied by protein or other non-lipid material.

d) Electron Microscopy of Membrane Thin Sections

In addition to these X-ray diffraction studies, electron microscopy of thin sections also allows direct visualization and comparison of membranes from a large number of different cells and organelles. Using a KMnO$_4$ fixation procedure, Robertson (1959, 1964) observed a three-layered unit, approximately 75 Å thick in myelin membranes, at the surface of many different cells and around many different cellular organelles. This unit appeared as two dense lines about 20 Å wide separated by a lighter, 35 Å wide space. ROBERTSON called this unique and consistently trilaminar structure a "unit membrane" (Fig. 2 c) and related this electron microscope image to the molecular models which had been provided by DANIELLI and DAVSON, SCHMIDT, SCHMITT, and FINEAN. ROBERTSON's unit membrane concept clarified and unified a body of diverse information and emphasized the relationship between the birefringence and X-ray diffraction work of the 1930's and the more recent electron microscope observations. ROBERTSON's model was, of course, very similar to its antecedent, the DANIELLI-DAVSON model. But, by referring to the biological membrane as a unit, ROBERTSON emphasized not only that all three layers of the 75 Å structure seen in the electron microscope were part of one membrane, but also that all membranes had a similar molecular arrangement and origin.

The characteristic 75 Å "railroad track" or unit membrane image may be seen in many different membranes following a variety of fixation and staining procedures (ELBERS 1964). The chemistry of many of these procedures remains to be studied although some progress has been made in elucidating the interactions between various membrane components and some of the most frequently used fixatives, OsO_4 and $KMnO_4$ (HAKE 1965, SCHIDLOVSKY 1965, STOECKENIUS and MAHR 1965). ROBERTSON (1959) assumed that the two black lines characteristic of the electron microscope image of an osmium-fixed membrane corresponded to stained protein. However, other evidence initially suggested that the dark lines represented the hydrocarbon portion of the lipid whose unsaturated fatty acids had been stained by reacting with the osmium fixative (STOECKENIUS 1959). This suggestion had important implications because the electron microscope image would certainly not support a PLP model if the outer black lines corresponded to lipid. However, since the staining patterns of a number of different membranes, including myelin (NAPOLITANO et al. 1967) and mitochondria (FLEISCHER et al. 1967), were unchanged by lipid extraction, it was subsequently concluded that the black lines must correspond to the protein components of the membrane.

It is probable that during OsO_4 fixation unsaturated fatty acids in the membrane are converted to stable glycol osmates (KORN 1967). Work with monolayers shows that the products formed by the reaction of osmium with unsaturated fatty acids are probably reoriented with the hydrophobic paraffin chain and attached osmium atom moved to the polar surface of the monolayer (DREHER et al. 1967). Although this osmium induced reorientation of unsaturated fatty acids has only been studied in monolayers, it may also occur in biological membranes. If so, extraction of the lipids should alter only the intensity and not the basic staining pattern of a PLP membrane. That this is precisely the case provides convincing evidence that the two dark lines of the electron microscope image of a single membrane must correspond to the location of a major portion of the protein in that membrane.

e) Lipid and Protein Measurements

The first suggestion that membrane lipids may be structurally arranged into bimolecular layers was made by Gorter and Grendel (1925) who extracted the lipid from a known number of erythrocytes and compared the area occupied by the lipid as a monomolecular layer with the known total surface area of the erythrocytes. Since the ratio of lipid area: membrane surface area was very close to 2 : 1 for erythrocytes from a number of mammals, it was concluded (see equation 4, above) that "the chromocytes are covered by a layer of fatty substance that is two molecules thick."

The basis of this ingenious experiment had two weaknesses: first, acetone was used as a lipid extractant, and a number of phospholipids are not readily soluble in pure acetone; secondly, the surface area chosen for erythrocytes was calculated from dried cells and, therefore, underestimated the true surface area. For instance, 99 μ^2 was used as a surface area for human erythrocytes, whereas more recent values range around 145 μ^2 (Westerman et al. 1961).

Bar et al. (1966) repeated these experiments using modern techniques for lipid extraction and a better estimate for erythrocyte surface area. When the extracted lipids were compressed in a Langmuir trough at low surface pressures (5–15 dynes per cm), a ratio of approximately 2 : 1 of monolayer area : cell surface area was obtained. This agreed with Gorter and Grendel's results. In Bar et al.'s experiments most, if not all, of the lipid was extracted. Apparently the incomplete lipid extraction of Gorter and Grendel was balanced by their low estimates of erythrocyte surface area.

The experimental approach outlined above may be compared with calculated areas for lipids in the same membrane system. Engelman (1969 a) has noted that the lipid composition of human erythrocyte membranes is known with reasonable accuracy so that it should be possible to calculate directly the area occupied by lipid within the membrane. Engelman first determined the total hydrophobic volume occupied by the lipid of a single erythrocyte membrane and then calculated the equivalent surface areas of phospholipid and cholesterol. He knew from X-ray diffraction studies of pure erythrocyte phospholipid-cholesterol systems that the packing of these lipid molecules would result in an area of 86 $Å^2$ for a phospholipid-cholesterol molecular pair (Rand and Luzzati 1968). However Engelman calculated that if these pairs were packed as a monolayer within the erythrocyte membrane, the area available for each pair would be only 58 $Å^2$, an impossibly small figure. On the other hand, if the lipids were packed as a bilayer, there would be 117 $Å^2$ available per phospholipid-cholesterol pair. Engelman reasoned that this was the more likely possibility and concluded that some of the space reflected by the difference between 86 $Å^2$ (for the lipid pair in the pure lipid system) and 117 $Å^2$ (for the calculated lipid pair area in the membrane) might be occupied by penetrations of protein into the lipid bilayer.

It is apparent that both surface chemistry and calculated values are consistent with the idea that lipid bilayers are present within erythrocyte membranes. We may ask if this is also true of other membranes which have lipid-protein ratios different from those of the erythrocyte membrane. For instance, is there enough lipid in mitochondrial inner membranes for a bilayer? This question could not be answered until recently since it was difficult to estimate surface area for complex membrane systems like mitochondria. However, now that we have reliable values for the lipid and protein content of relatively pure membrane preparations, the equations (1–3) discussed earlier may be used to give estimates of surface area. To do so, we simply calculate the area of one gram of membrane lipid-protein when it is spread at a given density to a thickness equivalent to that of the original membrane. To determine if there is enough lipid for a bilayer, we then assign molecular areas to each of the lipid components and calculate the area occupied by a monolayer of total lipid present in the gram of lipid-protein. The ratio of the lipid monolayer area to the area of the original membrane lipid-protein gives the number of lipid layers possible for that amount of lipid in the original membrane.

We have carried out such calculations for a number of membranes (Table 1). In doing so, the following considerations were necessary:

1. There is no firm upper limit to the area occupied by a lipid molecule in a membrane. However, the lower limits of lipid molecular areas have been established by X-ray diffraction studies. Therefore rather than guess at an area per lipid molecule in a membrane, we ask what area would be occupied by a tightly packed lipid bilayer in a given membrane, or conversely, how much free space is available within a membrane for material other than lipid. We chose 55 Å^2 for a phospholipid molecular area in membranes containing little or no cholesterol. This is an average molecular area for several phospholipids in hydrated bulk phase lipids (REISS-HUSSON 1967, LUZZATI 1968). On the other hand, we chose 50 Å^2 as a reasonable area for phospholipid in membranes containing appreciable quantities of cholesterol (RAND and LUZZATI 1968) because of the effect cholesterol has on molecular packing within phospholipid-cholesterol mixtures (DE BERNARD 1958, VAN DEENEN et al. 1962). The molecular area of cholesterol is constant and was taken to be 38 Å^2 (EKWALL et al. 1957).

2. A thickness of 80 Å was used as an average thickness for membranes in all the calculations. This approximates the best estimates of membrane thickness from electron microscopic data and is a typical value derived by X-ray diffraction studies of fresh, hydrated myelin membranes from several sources (WORTHINGTON and BLAUROCK 1969 a).

3. Since structural water contributes to the mass and volume of membranes, it is necessary to assume minimal hydration values. Hydration was taken to be 25 per cent in myelin and 15 per cent in other membranes (FINEAN et al. 1966). Density values for the hydrated membranes were taken from the literature or assumed to be 1.1 g cm^{-3} when no published values could be found.

A sample calculation is given below for the erythrocyte membrane:
1. Total lipid (per cent dry weight) = 40 per cent.
2. Water = 15 per cent.

Table 1. *Percent of Membrane Surface Area Covered by*

Membrane	References for Composition	Total Lipid Content	
		per cent dry weight	per cent wet weight
Myelin, human brain	O'Brien and Sampson 1965	78	58
	Norton et al. 1966	70	52.5
Microsomes	Rouser et al. 1968	53	45
Erythrocyte	Maddy and Malcom 1965	40	34
	Rouser et al. 1968		
Bacteria (*Streptococcus faecalis*)	Shockman et al. 1963	34	29
Sarcoplasmic reticulum	Deamer, unpublished results from lobster tail muscle microsomes	50	43
Mycoplasma laidlawii	Razin 1967	36	31
Mycoplasma mycoides	Razin 1967	40	34
Mitochondria			
Outer membrane	Thompson and Parsons 1970	45	38
Inner membrane	Thompson and Parsons 1970	22	19

a This is the ratio of the area of a monolayer of lipid derived from one gram membrane lipid-

3. Total lipid (per cent of hydrated membrane) = 34 per cent.
4. Cholesterol (per cent of total lipid by weight) = 27 per cent.

If we are concerned with 1 gram of hydrated membrane, then:

5. Total cholesterol weight = $0.27 \times 0.34 = 0.092$ g.
6. Molecular weight of cholesterol = 386.
7. Total area of cholesterol as a monolayer =
$$\frac{0.092 \text{ g}}{386 \text{ g mole}^{-1}} \times \text{Avogadro's number} \times 38 \text{ Å}^2 \times 10^{-20} \text{ m}^2 \text{ Å}^{-2} = 54 \text{ m}^2.$$
8. Total phospholipid weight = 0.34 g $- 0.092$ g $= 0.248$ g.
9. Average molecular weight of phospholipid = 800.
10. Total area of phospholipid as a monolayer =
$$\frac{0.248 \text{ g}}{800 \text{ g mole}^{-1}} \times \text{Avogadro's number} \times 50 \text{ Å}^2 \times 10^{-20} \text{ m}^2 \text{ Å}^{-2} = 93 \text{ m}^2.$$
11. Total area of lipid as a tightly packed monolayer = 93 m² + 54 m² = 147 m² or 74 m² as a bilayer.
12. Area of one gram of hydrated lipoprotein spread to a thickness of
$$80 \text{ Å} = \frac{1 \text{ g}}{80 \text{ Å} \times 1.13 \text{ g cm}^{-3}} = 110 \text{ m}^2$$
13. Per cent membrane area covered by a lipid bilayer = 74 m²/110 m² = 67 per cent.

Tightly Packed Lipid Bilayers in Various Membranes

Cholesterol per cent total lipid	Density	Area/Gram Lipid-protein	Area/ Monolayer	Ratio[a]	Per cent Surface Area of Membrane Covered by Lipid Bilayer
25	1.08	116	246	2.1	108
28	1.08	116	227	1.95	98
6	1.1	113	188	1.65	83
27	1.13	111	143	1.3	67
0	1.1	113	127	1.2	56
5	1.1	113	178	1.6	80
3	1.17	113	123	1.1	62
30	1.1	113	150	1.3	66
20	1.13	111	157	1.4	72
0	1.21	103	79	0.77	40

protein to the area of one gram lipid-protein spread to a thickness of 80 Å.

It is clear from the results presented in Table 1 that most membranes have enough lipid to cover a major portion of their surface with a tightly packed lipid bilayer. There are two important approximations used in the calculations. The first concerns the molecular areas of lipid molecules. The figures chosen represent minimal areas, and, if packing is looser in lipid bilayers than in lipid crystals, the area occupied by the lipid bilayer would be proportionally larger. The second approximation is the value for hydration of membranes. If in reality there is other than 15–25 per cent water in membranes, the relative area occupied by lipid bilayers would be altered. For instance, if water is ignored in the calculations for the erythrocyte membrane, the per cent surface area covered by lipid bilayer increases from 65 per cent to 75 per cent. However, the importance of including structural water in the calculations is clearly demonstrated in myelin. It is impossible to fit all the lipid into the myelin membrane if water of hydration is ignored.

These possible errors do not alter the final values by more than 10–15 per cent, and we can conclude with some certainty that most membranes have enough lipid to cover well over half their surface with lipid bilayer. It follows that the rest of the surface area is available to be occupied by material other than hydrocarbon chains of lipid. A similar conclusion was reached by ENGELMAN (1969 a) whose observations were described earlier. In light of freeze-etch and other data which will be discussed below, it seems reasonable to assume that the extra space within the lipid bilayer is occupied by functional and structural protein.

f) Thermal Analysis

Thermal analysis of membranes is based on the observation that when bulk phase phospholipids are heated, they show endothermic transitions which reflect marked changes in molecular organization and motion (Chapman and Wallach 1968, Ladbrooke and Chapman 1969). These phase transitions are shown by X-ray diffraction to involve primarily the hydrocarbon chains of the phospholipids (Luzzati 1968). At low temperatures the presence of sharp X-ray reflection around 4.2 Å indicates that the paraffin chains are ordered and stiff. The only structure compatible with stiff chains is lamellar (Luzzati 1968). As the temperature is raised, the sharp 4.2 Å reflection is replaced by a more diffuse band at 4.6 Å indicating that the phase transition involves a "melting" of the hydrocarbon chains. The melted chains may remain in a lamellar configuration in which the lipid molecules are liquid and exhibit a high degree of molecular motion, but there are other configurations which are compatible with liquid chains. These include hexagonal, cubic and other phases in which the hydrocarbon chains fill volumes of odd shape (Luzzati 1968).

When phospholipids contain short chains or unsaturated bonds, the phase transitions occur at lower temperatures than when they contain long or saturated chains. However, the transition temperature for different phospholipid classes may vary even when they contain the same fatty acid residues. When phospholipids are hydrated, the stiff to liquid transition usually occurs at a lower temperature than when they are anhydrous (Chapman et al. 1967).

The reversible thermal transitions involving chain melting depend upon the cooperative association between the hydrocarbon chains. This association is modified by factors which affect their organization or interactions. For example, the addition of cholesterol to synthetic phospholipids gradually decreases both the transition temperature and the amount of heat absorbed by the transition (Ladbrooke et al. 1968). In fact, no transition is observed in mixtures with equimolar ratios of lecithin to cholesterol. This sensitivity of the thermal transitions to the nature of the lipid interactions has formed the basis of several studies of membranes *in vivo*.

Steim et al. (1969) were the first to demonstrate thermal phase transitions in intact membranes. They choose *Mycoplasma laidlawii* for their studies because this organism has only one kind of membrane, the cell membrane, and its fatty acid composition can be readily controlled (Razin et al. 1967, McElhaney and Tourtellotte 1969). Furthermore, when grown on tryptose broth, the organism does not contain cholesterol (Razin et al. 1966 a, Razin et al. 1966 b, Steim 1968). This is important because, as noted above, phase transitions are hard to observe or are absent in the presence of cholesterol.

Fig. 5 shows the endothermic transitions of lipids, membranes and whole cells of *M. laidlawii* (Steim et al. 1969). Because the membrane transitions were reversible and occurred at the same temperature as transitions in the lipids extracted from these membranes, Steim et al. concluded that they arose from a change in the state of the lipids. Furthermore, Steim et al. reasoned that since the phase change is a melt involving the cooperative association between hydrocarbon chains, it would vanish or be perturbed by apolar binding of lipids to proteins. STEIM et al. concluded that the lipids in membranes are probably in a bilayer conformation with the hydrocarbon chains associated with each other rather than with proteins.

STEIM et al.'s results have been confirmed by MELCHIOR et al. (1970) who have used a very sensitive differential calorimeter to study *M. laidlawii* cells. By means of viable cell titers taken before and after cells were

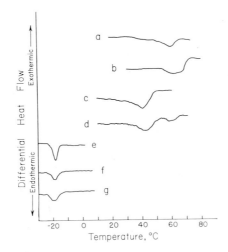

Fig. 5. Calorimeter scans of *M. laidlawii* lipids, membranes, and whole cells. *a*) Total membrane lipids from cells grown in tryptose with added stearate; *b*) membranes from stearate-supplemented tryptose; *c*) total membrane lipids from cells grown in unsupplemented tryptose; *d*) membranes from unsupplemented tryptose. The smaller transition at higher temperature was shown to be the irreversible denaturation of protein; *e*) total membrane lipids from cells grown in tryptose with added oleate; *f*) membranes from oleate-supplemented tryptose; *g*) whole cells from oleate-supplemented tryptose. The first four preparations were suspended in water; for the latter three scans, the solvent was 50% ethylene glycol containing 0.15 M NaCl. From STEIM et al. (1969).

submitted to calorimetry, these authors demonstrated unambiguously that the lipid phase transition is a property of living cells. These authors also examined membranes whose protein content was reduced by treatment with pronase. The pronase-treated cells underwent transitions similar to those observed in intact membranes and almost identical to those observed in membranes whose protein was previously denatured by heating. These results led MELCHIOR et al. to conclude that the conformations of most of the lipid and protein components of the membrane are independent of each other. From these and similar studies in which the phase transitions have been observed in the membranes of *E. coli*, *Micrococcus lysodeikticus* (STEIM 1970) and mitochondria (STEIM, personal communication) it is suggested that a major portion of the lipids of many membranes must be in the liquid-crystalline state characteristic of a bilayer.

The phase transitions in *Mycoplasma laidlawii* have also been studied by X-ray diffraction (ENGELMAN 1969 b). The results confirm that the transitions involve changes in the configuration of a lipid bilayer rather than changes between some of the other lipid phases described by Luzzati (1968). Below the transition temperature a sharp diffraction maximum near 4.2 Å indicated close hexagonal packing of relatively stiff fatty acyl chains. This packing is seen in phospholipid-water mixtures only when in a lamellar phase (Luzzati 1968). The low-angle diffraction patterns suggested that this same lamellar configuration existed above the transition temperature where the sharp 4.2 Å maximum was replaced by a broader 4.6 Å peak characteristic of more fluid fatty acyl chains. Thus the X-ray diffraction studies are in complete agreement with the calorimetric analyses of Steim et al. and Melchior et al.

g) Electron Spin Resonance

Electron spin resonance (ESR) techniques offer a method for determining the state of binding, local polarity, anisotropic motion and local viscosity of paramagnetic molecules in biological membranes. Because most biological systems are basically diamagnetic, paramagnetic labels must be introduced into the biological system to allow ESR analysis. Several approaches have been developed for introducing the paramagnetic "spin labels" into biological systems (Hamilton and McConnell 1968, Griffith and Waggonner 1969, Snipes and Keith 1970). The spin labels are usually stable, nitroxide free-radicals. Their chemistry is fairly well understood (Forrestor et al. 1968), and a wide choice of probes may be synthesized by known procedures. These spin labels may be adsorbed onto or absorbed into the membrane by non-covalent interactions which occur *in vitro* (Hubbel and McConnell 1968) or they may be incorporated, in some cases covalently bonded, into the membrane components by *in vivo* processes (Keith et al. 1968, Keith et al. 1970, Tourtellotte et al. 1970).

In a strong magnetic field the magnetic moment of the free-radical's unpaired electron is oriented either parallel or antiparallel to the field direction. These two orientations, called spin states, represent two different energy levels for the electron. Spin resonance occurs when the unpaired electron is promoted from the lower energy level to the upper level. The conditions for resonance and the precise shape of the absorption (resonance) lines give information about the orientation, local environment and molecular motion of the spin label.

In the first exploratory ESR studies of membranes the spin label 2, 2, 6, 6,-tetramethyl-piperidine-1-oxyl was dissolved in a variety of membranes including those of nerve fibers, erythrocytes, mitochondria and phospholipid vesicles (Hubbel and McConnell 1968). The results suggested that certain membrane systems, particularly those of the nerve, contain liquid-like regions of intermediate viscosity.

A more sophisticated approach to spin label studies was initiated by Keith et al. (1968), who utilized a specially synthesized nitroxide-stearate label (12 NS, Waggoner et al. 1969) which was incorporated into *Neurospora crassa* mitochondrial lipids *in vivo* and enzymatically bonded with glycerol moieties to form phospholipid molecules. The resonance spectra showed that the spin label was in a semiviscous, hydrophobic environment (Keith et al.

1970). This environment was contrasted with the one provided by mitochondrial structural protein hydrophobically associated with lipid. This hydrophobic association between protein and lipid—similar to that suggested in many LPL models—led to extreme immobilization of the probe. No such immobilization was ever observed in mitochondrial membrane preparations. Within the membrane the probe was always in an environment most similar to that found in aqueous dispersions of the total lipids extracted from mitochondria.

Similar results were obtained when a spin-labelled fatty acid was incorporated *in vivo* into the polar lipids of *Mycoplasma laidlawii* (TOURTELLOTTE et al. 1970). It was found that the ESR signal from either intact cells or their extracted lipids reflected the fatty acid composition of the *Mycoplasma* membranes. The composition of the *Mycoplasma* membrane could be varied by changing the growth medium. Comparison of signals from intact cells, gramicidin-treated cells, heat-treated cells and extracted lipids confirmed the existence of an extended region of apolar, lipid-lipid interaction in *Mycoplasma* membranes. The spin label was slightly but significantly less mobile in the intact membrane than in lipid extracts made from these membranes. This difference in probe mobility in the membrane and in the extracted lipids could be due to some weak lipid-protein interaction in the intact membrane. However, the differences could not be attributed to the type of strong hydrophobic lipid-protein interaction that occurs between lipids and mitochondrial structural protein. Similar results have been reported in a more recent study of *Mycoplasma* (ROTTEM et al. 1970).

Because the orientation of spin labels must reflect the local molecular structure, extremely interesting results have come from ESR experiments in which the orientation of the labels within membranes has been investigated. These experiments have utilized a variety of spin labelled fatty acid analogues in manually oriented nerve fibers and hydrodynamically oriented erythrocyte membranes (HUBBEL and McCONNEL 1969). In both systems, the preferred orientation of the spin labels was with the long hydrocarbon chain extended and perpendicular to the membrane surface. These studies also showed that motion of the spin labelled fatty acid analogue was greatest at the end of the molecule furthest from its polar end. Hence, the polar end of the molecule was important in anchoring the label. Although the motion of spin labels in dispersions of phospholipid bilayers, nerve membranes and erythrocyte membranes differed, in all cases the only plausible conclusion was that a substantial portion of all of these biological membranes contains a bilayer of lipid.

Conclusive as these observations may appear, it should be realized that they assume that the spin-labelled probe is not systematically excluded from environments and interactions in which unlabelled lipids participate. Even when the spin label is enzymatically incorporated into the phospholipids (KEITH et al. 1968, TOURTELLOTTE et al. 1970) it may not substitute for short chain or unsaturated fatty acids which are involved in important interactions whose physical parameters differ from those detected by the probe. Further ESR work using a greater variety of spin labels may provide evidence for such interactions.

h) Criticism of the Protein-Lipid-Protein Model

There are now obvious weaknesses of the PLP concept as a general model for all membranes:

1. The model, although structurally reasonable for relatively non-metabolic membranes like myelin, cannot account for all the known functional aspects of other membranes. A general model must explain differential permeability, active transport, excitability, oxidative phosphorylation and photophosphorylation. This difficulty has led to complete rejection of the PLP concept by some workers (GREEN and PERDUE 1966, KORN 1966) or more commonly to modification such as inclusions of functional protein within the structure of the membrane (LUCY 1964, WALLACH and GORDON 1968, ZAHLER 1969).

2. In the simplest PLP model, only electrostatic forces would bind protein to membranes. Although some proteins behave as though they were electrostatically bound to the membrane (HOOGEVEEN et al. 1970), a major portion of membrane proteins does not (ROSENBERG and GUIDOTTI 1969). Cytochrome c is an example of a specific protein removed from membranes by treatment designed to break electrostatic bonds (GREEN and FLEISCHER 1964), but many other membrane proteins require detergent treatment or drastic pH changes to bring about their solution (ROSENBERG and GUIDOTTI 1968 and 1969, RICHARDSON et al. 1963). This suggests that hydrophobic bonding may play an important role in stabilizing the structure of membrane lipid-protein.

It should be noted that there are two possible ways in which protein could be hydrophobically bound to a lipid bilayer without requiring the entire protein molecule to be located within the structure of the membrane. STOECKENIUS and ENGELMAN (1969) have suggested that apolar portions of the protein could in part penetrate to the hydrocarbon core of the lipid bilayer and thus be hydrophobically bound to the membrane. Alternatively, DEAMER (1970) has proposed a "lipid bridge" in which some of the membrane phospholipids would have one fatty acid chain extended into the membrane interior and the other fatty acid chain extended toward the membrane surface, where it would interact with hydrophobic binding sites of the surface protein. Such bridges have also been implied in accounting for the physical properties of model systems characterized by hydrophobic lipid-protein interaction (GULIK-KRZYWICKI et al. 1969).

2. Lipid-Protein-Lipid Membranes

The application of several relatively new physical techniques to the study of membrane structure has generated several conceptually related alternatives to the PLP model. Although it is difficult to provide a concise description of these alternative models, they do have certain features in common. First, large portions of membrane protein are considered to be in an apolar environment. Instead of protein being attached to continuous lipid bilayers as in the PLP model, a certain amount of the protein is considered to be within the internal structure of the membrane. Although lipid bilayers may be present, they are discontinuous, and the membrane may be considered a lipid-protein mosaic. As a corollary there is considerable interaction between membrane protein and the hydrocarbon chains of lipids, and a large number of the lipid head groups are directly exposed to the aqueous phase. Since much of the protein is within the structure of the membrane and many of the lipid head groups are on the surface of

Table 2. *Infrared Spectroscopy: Membrane Absorption Bands*

Band source	Frequency (cm^{-1})		
	Dry film	In H_2O	In D_2O
Proteins, Amide 1 [a]			
α-helix	1650	1652	1650
Random coil	1658	1656	1640
β-Pleated sheet	1632	1632	1632
Lipids [b]			
4 or more connected CH_2, all trans.	720		
phosphate P = 0 stretch	1225		

[a] From TIMASHEFF et al. 1967.
[b] From BELLAMY 1958.

the membrane, we will refer to this model as the lipid-protein-lipid or LPL model. In the discussion to follow, the evidence which has led to the LPL concept will be evaluated.

a) Infrared Spectroscopy

Infrared spectroscopy can distinguish between certain protein conformations and provide information about the organization of the lipid hydrocarbon chains in dried, hydrated or deuterated membranes. Table 2 shows some of the characteristic absorption bands in infrared spectra of membranes.

Infrared spectra of membranes were first investigated by MADDY and MALCOLM (1965, 1966) who examined dried films of hemoglobin-free erythrocyte ghosts. They found the amide I band at about 1648 cm^{-1} with no trace of a component at 1630 cm^{-1} and concluded that little β-pleated sheet conformation was present in membrane proteins of erythrocytes. Since they could produce β-conformation by denaturation, their findings show that there is nothing unique about membrane protein which prevents it from assuming the β-form.

MADDY and MALCOLM's conclusions have since been confirmed and extended by further infrared analysis of dry erythrocyte ghosts (CHAPMAN et al. 1968 c), fully hydrated fresh ghosts (MADDY and MALCOLM 1966, STEIM 1968) and other cell membranes (WALLACH and ZAHLER 1969). None of these studies found significant amounts of β-structure in the membrane proteins. However, mitochondrial membranes may be exceptional, as proteins extracted from them show a distinct shoulder near 1630 cm^{-1} (WALLACH et al. 1969). The general absence of β-form protein in membranes is not consistent with the supposition of an extensive array of β-conformation protein suggested in many PLP membrane models (VAN DEN HEUVAL 1963, KAVANAU 1965). However, the absence of β-conformation does not rule out other PLP structures. STEIM (1968) compared spectra of dry films and hydrated preparations of erythrocyte membranes both in H_2O and D_2O. For randomly coiled proteins, the amide I band is shifted down by about 10 cm^{-1} when H_2O is replaced by D_2O (see Table 2); upon complete exchange the amide II band around 1540 cm^{-1} disappears and is replaced by a band at about 1450 cm^{-1} (TIMASHEFF et al. 1967). STEIM found that when ghosts were transferred from H_2O to D_2O the amide I band moved from 1651 to about 1640 cm^{-1}, and simultaneously the amide II band at 1540 cm^{-1} diminished. He interpreted these changes as showing an appreciable amount of randomly coiled, water-accessible protein in the erythrocyte membrane. STEIM accounted for residual absorption at 1540 cm^{-1} by assuming the presence of some helical protein that neither exchanges with D_2O nor shifts its absorption peak. STEIM's results with erythrocyte membranes indicate that two-thirds to three-fourths of the protein amide groups are freely accessible to water and that much of the protein exists in an open, mostly random conformation. The results are consistent with both the PLP and LPL models.

Information about the lipids in membranes can also be obtained from infrared spectra, particularly from the absorption band at 720 cm^{-1} which has been assigned to the main methylene (CH_2) rocking mode (BELLAMY 1958). Because the 720 cm^{-1} absorption occurs when there are four or more CH_2-groups organized in an all-*trans* planar configuration, the presence or absence of the 720 cm^{-1} band can provide an indication of the lipid chain organization. Further indicators of lipid organization might be provided by the 1225 cm^{-1} band associated with phosphate groups.

CHAPMAN et al. (1968 c) have quantitatively compared the infrared spectra of dried erythrocyte membranes and their total extracted lipids. The spectrum of the extracted lipid was characteristic of phospholipids in a liquid-crystalline organization. The 1225 cm^{-1} frequencies associated with the phosphate groups were the same in both the erythrocyte and their total extracted lipids. On the other hand, the 720 cm^{-1} band was prominent in the spectra of the extracted lipids but extremely weak in the spectra of the original membranes. These results suggested that there was less all-*trans* planar character of the CH_2-groups of the lipid hydrocarbon chains within the membrane than in the lipids extracted from the membranes. This is apparently not the case for myelin where the spectra of both membranes and their extracted lipids show equivalent 720 cm^{-1} absorption bands (JENKINSON et al. 1969). CHAPMAN et al. explained the apparently greater disorder of the lipid chains in the membrane by assuming that they may adopt conformations other than all-*trans* because of hydrophobic associations with apolar amino acid residues of the membrane protein. This explanation supports the LPL concept. But an alternative explanation is simply that lipid is less tightly packed in the membrane. Looser packing would allow the lipid chains to flex and rotate away from the *trans*-configuration. This alternative explanation is in keeping with the PLP concept. Thus, infrared spectroscopy, while giving us some limits regarding individual lipid and protein configurations, does not provide enough information about lipid-protein interaction to distinguish between PLP and LPL concepts of the membrane.

b) Optical Rotatory Dispersion and Circular Dichroism

Because proteins have optically active absorption bands, their conformation may be studied by measurements of optical rotary dispersion (ORD) and circular dichroism (CD). Polypeptides in α-helical, β-, or random coil conformations exhibit distinctly different ORD and CD spectra. Mixtures of these conformations have CD and ORD spectra with intermediate band positions, shapes and amplitudes (WALLACH and GORDON 1967). It is therefore possible to deduce the conformation of unknown membrane proteins by comparing their spectra with those of synthetic polypeptides of known conformation. Although there are difficulties in determining the absolute amounts of α-helical, β- and random structure (GREENFIELD et al. 1967), useful limits can be set on the relative percentages of these different conformations.

The optical activity of various membranes, including eukaryotic plasma membranes, endoplasmic reticulum (WALLACH and ZAHLER 1966, WALLACH and GORDON 1968), erythrocyte ghosts (MADDY and MALCOLM 1965, LENARD and SINGER 1966, GLASER et al. 1970), mitochondrial membranes (STEIM and FLEISCHER 1967, URRY et al. 1967, WRIGGLESWORTH and PACKER 1968) and the plasma membranes of *Micrococcus lysodeikticus*, *Halobacterium halobium* and *Mycoplasma laidlawii* (STEIM 1968), has been studied. In all cases the ORD spectra exhibited a shape similar to that produced by an α-helix but of lower amplitude and with the entire spectrum shifted to longer wavelengths. These characteristics of the membrane ORD curves are reflected in the CD bands. The CD bands are characteristic of the α-helical spectrum but again are of lower amplitude and shifted to longer wavelengths than other helical proteins and polypeptides. It is generally agreed that these ORD and CD data, together with the infrared data cited above, show that a substantial amount of the membrane protein is in an α-helical conformation. Standard calculations assuming a mixture of α-helix and random coil estimate roughly 25 per cent α-helix (STEIM 1968), although a more extensive computer analysis suggests 50–60 per cent (WALLACH and GORDON 1968).

There is considerably less agreement regarding the source of the distortions and red shifts in the ORD and CD spectra. Because these unique characteristics of the spectra appeared to be associated with intact membranes, it was thought that these spectra represented some real but special property of membrane protein. WALLACH and ZAHLER (1966) and WALLACH and GORDON (1968) have attributed the red shift to interaction of protein with lipid, whereas Lenard and Singer (1966), Glaser et al. (1970), and Steim and Fleischer (1967) have suggested interaction of adjacent, aggregated protein helices.

There is some evidence which suggests that lipid-protein interaction is not the most probable explanation for the ORD and CD anomalies. For instance, Steim and Fleischer (1967) have shown that all of the characteristic optical features of intact membranes are found in lipid free preparations of aggregated membrane protein. Furthermore, protein-protein interactions may not be necessary to explain the distortions in membrane spectra if, as proposed by Urry and Ji (1968), these distortions are simply due to the particulate, light scattering nature of the system. Strong evidence supporting Urry and Ji's proposal comes from recent experiments which show that the distortions in the CD spectrum of erythrocyte membranes can be removed by reducing the scattering and that the characteristic red shifted membrane spectra can be reproduced by making a non-membranous helical protein act as a scatter (Schneider et al. 1970). Three types of experiments showed that the characteristic membrane spectra result from the superposition of a normal α-helical spectrum and an optical artifact. In the first experiment, the CD of intact erythrocyte ghosts was compared to that of sonicated ghosts. In the second, the CD of hemoglobin in solution was compared with that of hemoglobin as small scattering packets still inside the partially hemolyzed red blood cells. In the third, the CD of bovine serum albumin in solution was compared with that of the serum dispersed as small scattering droplets in silicone oil. The results of all three experiments showed that the distorted spectra characteristic of intact membranes cannot support a common and unique membrane protein conformation. Thus, whether membrane proteins are in hydrophobic interaction with lipids or in an aggregated state remains an open question which must be settled by techniques other than direct optical activity measurements of particulate, scattering membrane systems.

c) Nuclear Magnetic Resonance

Nuclear magnetic resonance spectroscopy (NMR) has recently been applied to the problem of molecular associations in biological membranes. The conditions for nuclear resonance are such that the line widths and other characteristics of radio frequency absorption allow one to investigate the motion of various components within hydrated, intact membranes in a magnetic field. The magnetic field experienced by the nuclei of any atoms within the membrane is produced by the magnetic fields of other nuclei of the same type as well as by the experimentally applied field. In solids, broad absorption lines are observed because chemically equivalent, neighboring, static protons with nuclear magnetic moments can cause variations in the strength of the local magnetic field experienced by any one nucleus. In liquids, narrow lines are seen because the magnetic interactions of equivalent neighboring nuclei can be considered as averaged out by their relative translational and rotational motion. Hence, if the period for overall rotation or reorientation of a component is long, *i.e.*, if the component is immobilized, narrow high resolution lines would not be expected. If, on the other hand, overall rotation or reorientation within the membrane is rapid, narrow high resolution lines might be expected.

This interpretation of line widths in terms of molecular mobility has been used by Chapman and his colleagues in their investigations of phospholipids and membranes. The usefulness of NMR was demonstrated by investigating molecular motion in various phospholipid phases (Penkett et al. 1968) and by studies of the molecular interaction between lecithin and cholesterol (Chapman and Penkett 1966). These investigations showed that narrow line, high-resolution NMR spectra of phospholipids dispersed in water were observed only after ultrasonic dispersion of the lipids. After sonic dispersion, the phospholipid is apparently in the form of small vesicles bordered by a lipid bilayer similar to that in the coarser, unsonicated aqueous phospholipid dispersions (Chapman et al. 1968 a). The increased resolution of NMR spectra of aqueous phospholipids after ultrasonication appears to be due to magnetic inhomogeneity effects which are present in coarse aggregates but are averaged out in the more rapidly spinning, smaller, sonicated preparations (Penkett et al. 1968).

Similarly, in intact erythrocyte ghosts or in homogenates of freeze-dried ghosts, well defined high resolution NMR spectra were obtained only after the preparations had been reduced to microsomal dimensions by sonication (Chapman et al. 1968 b). In the sonicated preparations, narrow line peaks were associated with $N(CH_3)_3$, $NCOCH_3$, and CH_2OC or CH_2OP groups of the membrane choline and sugar nuclei. These narrow line widths in the membrane spectrum were considered to represent segmental motion of the $N(CH_3)_3$, $NCOCH_3$ and CH_2OC or CH_2OP groups of the phospholipid molecules. It was reasoned that the motion of these groups reflected a low local viscosity in their immediate environment. However, the most important features of the membrane spectrum were the broadening of the $(CH_2)_n$ signal and the absence of high resolution signals from $CH=CH$ groups. The inhibition of the high resolution hydrocarbon chain signals was taken to indicate a high local viscosity around the hydrocarbon lipid chain in the membrane. Cholesterol-phospholipid interaction could not explain the signal broadening because spectra obtained from aqueous dispersions of the total lipids (including cholesterol) extracted from these red blood cell ghosts did not show similar inhibition of the $(CH_2)_n$ and $CH=CH$ peaks. Furthermore cholesterol extraction did not cause the $(CH_2)_n$ and $CH=CH$ signals to reappear. Thus, some feature of the membrane preparation produced line broadening and inhibited the appearance of the narrow hydrocarbon peaks.

Chapman et al. (1968 b) attributed the line broadening to lipid-protein interaction. According to this explanation, non-polar amino acids of the membrane protein interact with the hydrocarbon chains, increasing magnetic dipole interactions between adjacent CH_2 groups and restricting their segmental motion by raising the local viscosity. Chapman et al. concluded that the hydrocarbon chains of the lipid are partially interlocked with a portion of the membrane proteins. If correct, these conclusions provide strong support for the LPL concept.

Underlying CHAPMAN et al.'s conclusions is the assumption that molecular mobilities can be deduced from the reciprocal of line widths. This assumption may be valid for isotropic solutions of detergent micelles (CLIFFORD and PETHICA 1964) and serum lipoproteins (STEIM et al. 1968, CHAPMAN et al. 1969) but may not be justified in anisotropic particulate systems where there are a variety of line broadening mechanisms. Thus, for cell phospholipids or cell membranes dispersed in water, measured line width may not truly reflect the state of molecular mobility. For example, internal magnetic gradients caused by local variations in diamagnetic susceptibility may produce magnetic field-dependent broadening in phospholipids in water (SHEARD 1969). The broadening is more accentuated for the absorption lines of the fatty acid methylene protons than for the polar ends of the molecule.

In order to eliminate spurious line broadening mechanisms and to obtain a better estimate of molecular motion KAUFMAN et al. (1970) examined intact, unsonicated erythrocyte ghosts and their extracted lipids by two pulsed NMR techniques. The first, a free induction decay method, provides a measure of molecular mobility which is subject to the same kinds of error as calculations based on line width measurements. The second, a modified spin-echo method, provides a measure of molecular mobility which is independent of field inhomogeneities. When estimated by the first method, line broadening due to field inhomogeneities occurred in both membranes and their extracted lipids. When estimated by the second method, the true measure of molecular mobility in the membrane gave no evidence of extensive lipid-protein interaction. These results show that broad lines obtained by absorption NMR are misleading when interpreted in terms of molecular motion. Indeed, it was suggested that molecular mobility of the lipids was high and approximately the same in both membranes and aqueous dispersions of their extracted lipids. Thus NMR techniques have not established convincing evidence for the extensive hydrophobic lipid-protein interaction suggested by many LPL models.

d) Enzymatic Hydrolysis of Membrane Components

If lipid head groups are exposed on surfaces of membranes, it should be possible to hydrolyze enzymatically any of several ester linkages which compose the head group. This was tested by LENARD and SINGER (1968), who treated erythrocyte membranes with phospholipase c which catalyzes the hydrolysis of phospholipids to diglycerides and water soluble phosphorylated amines. Indeed, phospholipase c released up to 74 per cent of the phosphorous from erythrocyte membranes but apparently left the membranes intact and had little or no effect on the protein CD spectra (GLASER et al. 1970). GLASER and his colleagues concluded that the phosphoester bonds are readily accessible to the phospholipase c molecule in the intact membrane and that electrostatic interactions between phospholipid and membrane proteins play only a secondary role in maintaining the integrity of the mebranes. In fact, they interpret their results as evidence for the independence of membrane lipid and protein, the one having no influence on the molecular configuration of the other. MARTONOSI (1968 a) treated sarcoplasmic reticulum vesicles with phospholipase c and found that the hydrolysis of lipids had little effect on the appearance of the membranous vesicles, but he noted some shrinkage and nodule production. Similar nodule formation occurs when red cell ghosts are treated with phospholipase c (OTTOLENGHI and BOWMAN 1970, COLEMAN et al. 1970). Even if these nodules represent the diglycerides released from the remaining protein, the phospholipase c experiments of GLASER et al. (1970) suggest that the chemical structure and physical state of a large portion of the phospholipids can be radically altered with little detectable effect on the average conformation of the membrane protein. The accessibility of membrane lipids has also been demonstrated by OTTOLENGHI and BOWMAN (1970) and by COLEMAN et al. (1970) using phospholipase c on mitochondria and erythrocyte ghosts and by YU et al. (1968) who used phospholipase d on muscle microsomes. YU et al. noted that phospholipase d readily attacked the microsome membranes without inhibiting their calcium binding activity.

Evidence that lipases can attack membrane lipids is consistent with the idea that lipid head groups are exposed at the membrane surface, but other interpretations are also possible. In the first place, it appears that the susceptibility of any membrane component to lipase attack is controlled by a number of factors which may have little or nothing to do with its location in or on the membrane. Thus the cell membranes of isolated *Avena* coleoptile protoplasts are resistent to a wide range of both lipases and proteases (RUESINK and THIMANN 1965), whereas the membranes of lobster muscle fibers are sensitive to both lipases and proteases (GAINER 1967). In the second place, it is evident that membrane lipids exchange with non-membrane lipids (WIRTZ and ZILVERSMIT 1968) and are continuously being plugged into and out of the membrane matrix. It is not clear where in this process the phospholipase enzymes are actually hydrolyzing the lipids. Finally, there is no evidence that the lipases could not penetrate part or all of the membrane to attack the membrane lipids. The very fact that over 70 per cent of the membrane phospholipids are hydrolyzed by phospholipase c (LENARD and SINGER 1968) is in itself an inidication that precisely such penetration may occur. The LPL models usually assume that about half the lipid polar groups are exposed on each surface (LENARD and SINGER 1966, WALLACH and ZAHLER 1966). If the lipase were unable to penetrate the membrane protein, it would hydrolyze only the lipid exposed on the outer surface of an intact, closed membrane vesicle, and less than 50 per cent of the lipid phosphorus should be removed. Thus although it is indeed noteworthy that membranes are able to remain intact even after hydrolysis of 70 per cent of their phospholipid phosphoric acid residues, the enzymatic suceptibility of these groups may simply be a measure of the relative mobility and permeability of a dynamic protein meshwork, rather than an assay of lipid localization.

One way to check the above suggestion would be to stabilize membrane structure before carrying out enzymatic hydrolysis. If lipid head groups are actually at the surface of a membrane, they should still be susceptible to enzymatic attack after the membrane structure has been fixed. MICHELSON and DEAMER (in preparation) have fixed erythrocyte ghosts with glutaraldehyde and compared the action of phospholipase c on the fixed and unfixed control ghosts. Choline, phospholipids and cholesterol were still extractable from the fixed membranes, showing that glutaraldehyde does not cross-link these lipids to other membrane components. Nevertheless, phospholipase c attacked the lipids of the fixed membranes much less rapidly. It was concluded that lipid head groups were partially protected by a protein layer, but that in unfixed membranes the dynamic state of the protein molecules permitted enzyme molecules to react with lipid head groups. According to this interpretation, experiments based on enzymatic hydrolysis of membrane components are far from having proven the localization of lipid head groups at the membrane surface.

e) Criticism of the Lipid-Protein-Lipid Model

The LPL model, as a concept, has several strengths. It is consistent with our current understanding that hydrophobic bonds play a key role in determining the specific conformation of proteins (PERUTZ 1965) and thus offers a means by which membrane lipid-protein interactions may be stabilized in an aqueous environment. Furthermore, it rationalizes the data about solubility properties of membrane components. The observations that membrane proteins are often insoluble in aqueous phases unless detergents are present and that membrane lipids may be extracted by organic solvents support the idea of extensive hydrophobic interactions between membrane protein and lipid.

However, evidence for the LPL model is not satisfactory. As discussed above, there is no data available which clearly demonstrates extensive LPL configurations in any membrane. We conclude that there may be protein contained within the internal structure of some membranes and that lipid polar groups may be exposed occasionally at membrane surfaces but that the LPL structure cannot be considered an adequate general or specific model for membrane organization.

3. Particulate Membranes

All biological membranes are obviously particulate in that they have large quantities of protein involved in their structure, and single proteins are large enough to be visualized as individual particles by electron microscopy. It is not in this sense that we are here defining particulate membranes. Rather, a particulate or subunit membrane is defined as a two dimensional array of monomers, each monomer being a lipid-protein complex which has a specialized function and structure. According to some proponents of this hypothesis (GREEN et al. 1967), membranes may be disassembled into their subunits, and these subunits may in turn be reassembled into functioning membranes. The monomers are supposed to be held together in the polymeric membrane by hydrophobic bonding between proteins, and the lipids within one subunit are also presumed to be hydrophobically bound to the protein monomer. The particulate membrane hypothesis may be viewed as a variation of the LPL concept, since the lipid polar groups would be on the outer surface of the subunits with the hydrocarbon tails projecting inward.

The particulate membrane hypothesis imposes a greater order on membrane structure than has heretofore been suggested. Rather than having an indefinite and relatively random array of lipid and protein molecules within a membrane, the particulate membrane hypothesis suggests that lipids and proteins are ordered into quite defined functional subunits which remain intact even after a membrane is solubilized by detergents. The major questions at present are whether such order is justified by experimental evidence and whether it is necessary to explain known membrane functions. One of the concept's most attractive features is that it can account for membrane biogenesis as the synthesis of specific monomers which later polymerize into functioning membranes. The chief failing of the particulate membrane hypothesis is the lack of convincing evidence that all membranes or even some are composed of extensive complexes of this nature.

a) Chloroplast Membranes

Evidence supporting the subunit hypothesis has come from studies of several membranes, but there are few which have been as actively studied as the photosynthetic lamellae. Many studies suggest that these unique membranes are composed of a two dimensional array of subunits, and this evidence has in turn been generalized to suggest that all membranes are composed of subunits. A critical examination of photosynthetic membrane structure is therefore required to determine whether the evidence regarding their subunit character is sound and whether the fundamental architectural scheme of these lamellae is similar to that of other membranes.

Some of the earliest suggestions regarding the structure of photosynthetic lamellae were made by Frey-Wyssling and Steinmann (1948), who investigated chloroplast birefringence and dichroism. They found that the intrinsic birefringence of the lamellae was weak and therefore proposed that the lipids were not highly ordered. However, Frey-Wyssling and Steinmann used OsO_4 fixation to avoid loss of chloroplast lipids. Since OsO_4 is now known to alter the orientation of the lipids (Dreher et al. 1967), their conclusions apply only to fixed membranes and not necessarily to the membranes in vivo. Nonetheless, Goedheer (1957), Sauer and Calvin (1962), and Olson et al. (1962, 1964, 1966) have confirmed part of Frey-Wyssling and Steinmann's conclusions by measuring pigment dichroism in unfixed photosynthetic lamellae. Only a small percentage of the chlorophyll and carotenoids in the photosynthetic lamellae was oriented within the membrane. However, these pigments represent only half of the membrane lipids. Other lipids could be oriented to form a matrix around pigment-rich photosynthetic centers. Thus optical studies do not rule out the possibility of a PLP membrane in which a lipid bilayer, interrupted by numerous pigment or pigment-protein complexes, forms the matrix of the membrane.

X-ray diffraction techniques have been applied to chloroplast membranes by Kreutz (1966, 1969), Menke (1962), and their colleagues. They found that the scatter diagrams from fresh plastids could not be interpreted exclusively as indicating lamellar layering. Kreutz has explained this scattering as resulting from a two-dimensional array of subunits which forms crystallites in the protein layer of the membrane. However, the elegant work performed by Kreutz and his colleagues in deriving density distribution functions from X-ray scatter intensities is ultimately restricted by the interpretation of the density functions in terms of a concrete model. Kreutz himself acknowledges that the validity of this model can only be demonstrated by independent experimental manipulations, including electron microscopy (Kreutz 1966).

Some of the earliest electron micrographs of biological structures were those of photosynthetic membranes prepared by drying isolated chloroplasts on an electron microscope grid (Kausche and Ruska 1940). In order to bring out more detail, Steinmann (1952) shadowed such preparations and found that the surfaces of these lamellae were covered with particles about 200 Å apart. Frey-Wyssling and Steinmann (1953) interpreted these electron micrographs as showing closed flattened disks (thylakoids) made up of subunits. These observations were confirmed by Park and Pon (1961, 1963) who localized the particles on the interior surface of the thylakoid and by Park and Biggins (1964) who noted that these particles sometimes existed in very highly ordered arrays. In these arrays, the particles had dimensions of $155 \times 180 \times 100$ Å and appeared to consist of several subunits. Assuming that the membrane components were uniformly distributed over the lamellae, Park and Biggins (1964) estimated the chemical composition of a single $155 \times 180 \times 100$ Å particle on the basis of known chloroplast composition. Park (1962) concluded that the particles could be the morphological expression of what Arnold (1933) had termed a photosynthetic unit. Because this concept of a photosynthetic unit was conceived of in terms of the minimum number of chlorophyll molecules needed to fix one molecule of carbon dioxide per flash of light (Emerson and Arnold 1932), Park termed the supposed morphological photosynthetic unit a quantasome. However, the functional photosynthetic unit may be larger than one quantasome (Izawa and Good 1965), and individual units with the size and function attributed to the quantasome have never been isolated.

When thin sections of photosynthetic lamellae, fixed and stained with heavy metals, are examined in the electron microscope, the thylakoid membranes may show a tripartite structure (Mühlethaler 1960) similar to the unit membrane described by Robertson (1959, 1964). However, numerous observations of such sections (for review see Branton 1968) have shown the presence of transverse densities which have been the basis for elaborate speculations and drawings that assume a globular subunit structure in the lamellae (Benson and Singer 1965, Weier and Benson 1967). According to Weier and Benson (1967), the subunits appear to be approximately spherical with light cores and dark rims. The light cores averaged 37 Å in diameter and the dark rims were about 28 Å wide. Thus, the entire subunit was about

90 Å in diameter. It was suggested that these subunits, with groups of four being the possible equivalent of one quantasome, form lamellar aggregates which are the thylakoid membranes.

The suggestion that membrane subunits can be seen in cross sectioned thylakoid lamellae requires careful scrutiny, particularly since similar claims have been advanced in connection with a number of different membrane systems (ROBERTSON 1966). Sections for electron microscopy are usually thicker than 400 Å, and an electron microscope would not resolve one layer of particles but at best produce a compound image of several particles. A reasonable explanation of the cross densities which appear as subunits is that certain spots on the membrane surface are more densely staining than others. Because these spots react with and accumulate large amounts of stain (osmium, permanganate or lead), the stained materials on the two surfaces of the membrane appear to be confluent, giving the impression of a transverse density through the center of the membrane. Thus, predominant transverse densities may be due to a mosaic on the membrane surface, and currently available thin sectioning techniques do not provide clear evidence either for or against the existence of morphological membrane subunits.

When isolated chloroplast lamellae are negatively stained, numerous particles, 75–110 Å in diameter, are frequently found associated with the membranes. These particles have been interpreted as membrane subunits (ODA and HUZISIGE 1963), but subsequent work has demonstrated that most of the particles seen in such negatively stained chloroplast lamellae are not actually an intimate part of the membrane structure (HOWELL and MOUDRIANAKIS 1967, PARK and PFEIFHOFER 1968, ARNTZEN et al. 1969). For example, one type of particle which has been identified as a membrane component (MÜHLETHALER et al. 1965, ODA and HUZISIGE 1965) is morphologically identical to carboxydismutase (HASELKORN et al. 1965, TROWN 1965), one of the major proteins of the chloroplast matrix. Another class of particles has been correlated with the red protein, rubimedin (HENNINGER et al. 1966), whose specificity in the electron transport system has yet to be demonstrated (HENNINGER and CRANE 1966). Both classes of particles can be removed from chloroplast lamellae by washing in dilute buffer (HENNINGER et al. 1966, VON WETTSTEIN 1966, BRANTON and PARK 1967); the washed membranes show little particulate structure. The particles which remain belong to yet another class of proteins and contain much of the ATPase activity of the membranes (MOUDRIANAKIS et al. 1968, ARNTZEN et al. 1969). These ATPases are removable by washing in EDTA (HOWELL and MOUDRIANAKIS 1967, PARK and PFEIFHOFER 1968, ARNTZEN et al. 1969).

b) Other Membranes

A particulate membrane concept was first advanced by SJOSTRAND (1963) to explain certain electron microscopic images of mitochondria and other cytomembranes in which the membrane appeared discontinuous as it does in some chloroplast micrographs. SJOSTRAND interpreted the discontinuities as representing a two dimensional array of lipid globules or globular protein molecules. Sjostrand and Barajas (1969) have provided further evidence that particulate structures are present within mitochondrial inner membranes and have also criticised the trilaminar image of membranes which has been used as evidence in support of PLP membrane models. They reasoned that the dehydration and curing procedures of standard electron microscope techniques very probably denature membrane protein. Rearrangement of the denatured protein might result in artifactual images. Therefore, a procedure of glutaraldehyde fixation followed by ethylene glycol dehydration was devised to provide minimal denaturing conditions. With this technique, Sjostrand and Barajas demonstrated discontinuous images of mitochondrial membranes and interpreted these as particulate arrays of protein within the membranes. The classical trilaminar image was produced only when denaturing conditions (heat or acetone dehydration) were introduced during specimen preparation. It was concluded that the trilaminar image is an artifact produced by rearrangement of denatured protein during specimen preparation.

Lucy (1964) and Lucy and Glauert (1963) have proposed a dynamic membrane model in which the lipid may undergo a transition from bilayer configurations to a micellar state. In the micellar state proposed by Lucy and Glauert, micellar lipid and globular protein are present in a mosaic of 40 Å subunits. In this sense their model may characterize a particulate membrane structure. Such a model provides for a greater range of structure-function interactions for membrane lipid than do either the PLP or LPL models. For instance, water filled, 8 Å diameter pores are an intrinsic feature of the proposed structure. These pores could explain the high permeability of some membranes to water. A second advantage of the micellar membrane model is that selective permeability to cations is more readily explained than in bilayer models since charged pores are an intrinsic feature of the micellar phase. However, this interesting model so far is almost entirely speculative, and little direct evidence is available to support the possibility of micellar structures in membranes.

The particulate membrane concept has been most vigorously extended and amplified by Green and his co-workers (1967). Their original arguments were based on the observation that particulate structures protrude from negatively-stained mitochondrial inner membranes (Fernandez-Moran et al. 1964). It was suggested that these particles were repeating subunits which represented the entire electron transport unit of the inner membrane (Fernandez-Moran et al. 1964). However, Kagawa and Racker (1966) have since shown convincingly that the particles contain only the ATPase activity associated with inner membrane preparations. Furthermore, the particles are soluble in aqueous phases and contain no lipid. To accommodate these findings within the general concept of a particulate membrane structure, Green and Perdue (1966) have proposed an elaborate "tripartite" subunit, containing a head group (the ATPase), a stalk and a base piece. It is proposed that only the base piece is involved in the actual hydrophobic interactions which permit the tripartite subunits to associate into membranes.

c) Disassembly—Reconstitution Experiments

GREEN and collaborators (1967) generalized from the particles of mitochondria and suggested that many membranes are composed of particulate subunits. They reasoned that if membranes are particulate, it should be possible to break them down into monomers and then repolymerize them. This has been a major effort during the past several years with some

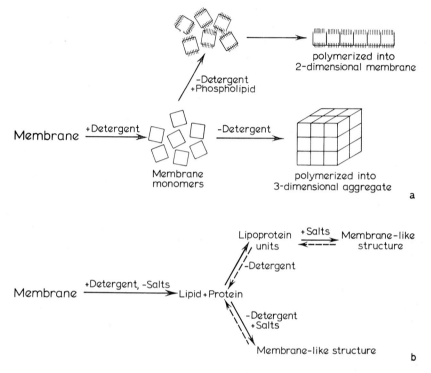

Fig. 6. Models for detergent solubilization of membranes. *a)* According to the subunit hypothesis, detergents permit a reversible dispersion of polymeric membranes into their component monomers. From GREEN and KOPACZYK (1966). *b)* An alternative explanation, supported experimentally (ENGELMANN and MOROWITZ 1968 a), is that detergents disperse continuous lipid-protein membranes into separate lipid and protein fractions which can re-aggregate into membrane-like structures.

apparent success. In a typical reconstitution experiment, membranes are first treated with a detergent, for instance bile salts, which causes disaggregation of the membrane. If the bile salts are then removed by dialysis, large three dimensional aggregates appear in the solution. However, if phospholipid is added, removal of the bile salts results in the formation of membrane vesicles. This experiment has been carried out on a wide variety of membranes (GREEN et al. 1967), and it was found that membranous vesicles were always observed following removal of detergent molecules in the presence of lipid. GREEN et al. interpret their results as shown in Fig. 6 a. According to this interpretation detergents cause the subunits within the original membrane to disassociate into monomeric forms. Upon removal of the detergent in the presence of added phospholipid, the monomers reassemble into the original membrane structure. Phospholipid inhibits three dimensional aggregation, perhaps by binding to the repeating units in such a manner that the units interact hydrophobically in only two dimensions as shown in Fig. 6 b.

Membrane formation from solubilized membranes or from isolated membrane derived enzymes has been repeated in many laboratories. Terry et al. (1967), Engelman and Morowitz (1968 a) and Razin et al. (1969) used sodium dodecyl sulfate (SDS) to solubilize *Mycoplasma* membranes. Upon removal of the detergent, membranous structures reappeared. Yanagida and Noda (1969) utilized a number of treatments, particularly high pH, to solubilize plasma membranes of myxamoebae. At pH 11, the membranes dissociated into particulate structures with a 7 S sedimentation coefficient. Lowering the pH caused a reaggregation of the particles into several morphological forms, among them structures resembling the parent membrane, double-thickness membranes, filaments, and random aggregates. Crane et al. (1968) combined cytochrome oxidase with phospholipid and produced extensive membranous structures which appear to consist of aggregated 50 Å particles, presumably cytochrome oxidase, embedded in a lipid bilayer and Stoeckenius (1970) mixed cytochrome c with mitochondrial phospholipid to produce membranous vesicles.

d) Criticism of the Particulate Membrane Hypothesis

The studies described above have established that lipid-protein membranes may be formed from mixtures of lipid and membrane derived proteins and provide most of the evidence used to support the particulate membrane hypothesis. However, these conclusions assume that if detergent solubilization produces subunits of uniform size, the subunits form membranes of the same composition and organization as the original membrane. A serious criticism of this assumption has been raised by Engelman and Morowitz (1968 b) who investigated detergent solubilization and reassociation of *Mycoplasma* membranes.

In preliminary experiments, Razin et al. (1965) found that SDS solutions readily solubilized *Mycoplasma* membranes. Furthermore, the solubilized component displayed a single schlieren peak, suggesting that *Mycoplasma* membranes may indeed be composed of lipid-protein subunits. However, in later studies it was found that the SDS-solubilized membrane was actually composed of two components, one consisting of lipid and the other of protein. Engelman and Morowitz (1968 b) interpreted this result as shown in Fig. 6 b. Detergent treatment breaks the membrane down to lipid and protein; it does not simply disperse subunits which existed in the original membrane. Removal of the detergent results in membrane formation, but not necessarily by polymerization of preexisting subunits. Indeed, lipid and protein rather than lipid-protein subunits may recombine directly to form a two-dimensional, membrane-like structure. It is apparent that membrane solubilization and reconstitution experiments can be readily interpreted in terms of the PLP concept and do not imply that structural subunits are present within the original membrane.

In this light, it is interesting to note that GENT et al. (1964) found that lysolecithin caused myelin to break down into lipid-protein subunits of uniform size, a result which proponents of the particulate model would construe as evidence for subunit structure of the original membrane. Indeed, GENT et al. concluded on the basis of this evidence that myelin had a subunit structure. However, myelin is the one membrane which is amenable to relatively definitive X-ray diffraction studies. Its structure has been clearly shown to be lamellar, with no evidence of subunit structure (CASPAR and KIRSCHNER 1971).

STOECKENIUS and ENGELMAN (1969) and STOECKENIUS (1970) also questioned the assumptions of the evidence, particularly the electron microscopic data, used to support the particulate membrane concept. STOECKENIUS (1970) notes that mitochondria may appear either particulate or trilaminar depending on the preparation. For instance, particles may be seen on "weakly stained" membranes, whereas they disappear in more heavily stained membranes. Particles may also be produced by aggregation of the staining metals themselves under electron bombardment in the microscope.

STOECKENIUS (1970) also criticized the evidence derived from disruption reconstitution experiments. He repeated the experiments in which mitochondrial membranes were dissolved in detergents and reconstituted by removal of the detergent. He found that the membranes were not, in fact, completely dispersed by the detergent but that many smaller pieces recognizable as membranes were present. STOECKENIUS concluded that the reconstitution evidence could be interpreted as reformation of membranes from small pieces of membranes rather than from solubilized subunits as visualized by GREEN and his co-workers (1967).

In summary, some membranes contain suggestions of a subunit structure, but there is no evidence for a subunit hypothesis that cannot be reconciled with a PLP model. Proof of this hypothesis will depend upon the isolation of the putative subunits and the demonstration that these subunits existed as such in the intact membrane. This has not been done.

4. Conclusions about Membrane Models

Results from the variety of approaches and physical probes that have been used to study membrane structure show that the conformation of the hydrophobic portions of membrane lipids is not strongly dependent upon the presence or arrangement of membrane proteins. Thermal analysis, ESR, NMR, and X-ray diffraction all suggest that the bulk of the membrane lipids behaves as though it were in a fluid bilayer whose conformation is chiefly determined by lipid-lipid interactions. Only a PLP model provides an organization which can reasonably account for the X-ray diffraction results with a variety of membranes as well as the detailed electron density profiles of myelin. Furthermore, the assignment of protein and lipid implicit in the PLP model is supported by electron microscopy.

In spite of the small differences between the conformations and mobilities of the membrane lipids *in situ* and of the extracted lipids in bilayers, there is certainly no good evidence for the strong lipid-protein interaction envisaged by both the PLP and the subunit models. Infrared data showing little or no β-conformation protein can be accommodated by a modified Danielli model in which the surface proteins are in random coil and a-helical conformations. Explanations of ORD and CD data which suggest hydrophobic lipid-protein interactions neglect light scattering effects which may interfere with optical rotation measurements. Interpretations of NMR line widths in terms of lipid hydrocarbon immobilization neglect field inhomogeneities and other line broadening factors which may operate in anisotropic systems. Enzymatic extraction techniques purporting to show the surface location of lipid moieties fail to take into account the mobility and turnover of membrane components. Finally, attempts to isolate a membrane subunit have produced lipid-protein aggregates which can as easily be interpreted as creations of the isolation procedure as actual components of the original membrane.

III. Differentiation and Specialization of Membranes

Although this summary of the current evidence strongly supports the PLP model, it does not show how a lipid bilayer with protein coated surfaces could account for the functional attributes of a membrane. Indeed, some modification and differentiation of the overall PLP design must be envisaged to explain the specialized permeability and functional properties of membranes, and regions of differentiation may involve the hydrophobic lipid-protein interaction suggested by many LPL models. However, the existence of such regions does not imply that the entire membrane is so organized. The available evidence leads us to visualize instead a lipid bilayer interrupted by localized intercalations of other membrane components. These intercalations may be protein or even lipid-protein. However, the number of such intercalations needed to explain membrane function may occupy only a limited portion of the membrane surface area.

There are in fact two systems in which it is possible to estimate the surface area occupied by a functional membrane protein. One of these, sarcoplasmic reticulum, has a highly active, ATP-dependent calcium transport system which is apparently incorporated into the membrane structure (EBASHI 1958). ATPase activity and calcium uptake are preserved in muscle microsome preparations, and it is therefore possible to estimate what fraction of microsomal membrane protein corresponds to the ATPase activity.

The ATPase of muscle microsomes has an estimated molecular weight of 175,000 (VEGH et al. 1968). Although turnover numbers for this ATPase are not available, turnover numbers for other microsomal ATPases have been estimated and range from 8,000–16,000 (BADER et al. 1968). ATPase activity associated with calcium transport ranges from 3×10^{-3} (mammalian) to 20×10^{-3} moles (crustacean) ATP hydrolysed per minute per g protein (MARTONOSI 1968b; D. W. DEAMER, unpublished results). We can calculate the amount of membrane protein which represents ATPase in lobster muscle microsomes as follows:

1. Number of ATP molecules hydrolysed per minute per g protein =
   20×10^{-3} moles ATP min^{-1} g^{-1} × Avogadro's number = 120×10^{20}.

2. Number of ATPase molecules per g protein =
   $$\frac{120 \times 10^{20} \text{ molecules min}^{-1} \text{ g}^{-1}}{8000 \text{ molecules/ATPase min}^{-1}} = 15 \times 10^{17} \text{ ATPase molecule g}^{-1}.$$

3. Weight of ATPase per g protein =
   $$\frac{15 \times 10^{17}}{\text{Avogadro's number}} \times 1.75 \times 10^5 = 0.44 \text{ g, or 44 per cent of the protein.}$$

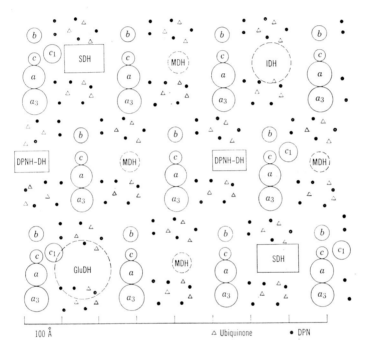

Fig. 7. Packing density of respiratory enzymes in mitochondrial inner membranes. From KLINGENBERG (1968).

This result is a maximum value. If 16,000 is used as a turnover number, the ATPase would correspond to 22 per cent of the membrane protein. Similar calculations for mammalian muscle microsome preparations, which may not be as pure, give values of 5–10 per cent of the membrane protein as ATPase (BADER et al. 1968).

A second membrane structure in which the amount of functional membrane protein can be estimated is the inner mitochondrial membrane. The electron transport enzymes of the respiratory chain have been thoroughly studied, and values for their molecular weight and relative abundance in the membrane are available. (For review, see KLINGENBERG 1968.) In general, the respiratory enzymes compose about 20–25 per cent of the total inner membrane protein although this value may rise to 50 per cent in washed electron transport particles (KLINGENBERG 1968). KLINGENBERG has estimated that approximately 10 per cent of the inner membrane surface area would be occupied by the respiratory components. In Fig. 7 the enzymes are drawn as they might appear within the membrane. It is apparent that even in these highly metabolic membranes only a fraction of the surface area is necessarily occupied by functional protein.

A. Differentiation of Boundary and Enzyme Functions

The idea that differentiation may occur within the plane of the membrane is certainly not new, and the growing realization that specific enzymatic properties are associated with dienerent membrane types has led to the suggestion that there may be enzymatically active proteins in addition to simple pores in the membrane matrix (see Fig. 2 b). MITCHELL (1957, 1961, and 1963) and MITCHELL and MOYLE (1958) have argued elegantly that the spatial anisotropy of the membrane provides a vectorial character to what might otherwise be scalar processes. This argument is based on the notion that the biological membrane separates an inside space from an outside space. If an enzyme, whose role is that of group transfer, exists within the membrane matrix, it could transfer groups across the membrane from the outside environment to the inside environment. For example, metabolic enzymes, properly located within the hydrophobic matrix of the membrane, might mediate transport, translocate protons, or couple metabolic oxidation to phosphorylation (MITCHELL 1967). Bacterial systems have provided some of the best evidence for the presence of vectorial enzymes which, more than simply being attached to the membrane, utilize the barrier properties of the membrane in carrying out their own functions. A series of functionally specific and inducible enzyme systems which mediate the transport of sugars, amino acids and inorganic ions have been studied in bacteria (PARDEE 1968, ROSEMAN 1969, KABACK 1970, KENNEDY 1971). Although a complete discussion of these systems is beyond the scope of this review, it has become clear that the bacterial cell membrane is a heterogeneous structure in which boundary functions and enzymatic functions may be ascribed to different portions of the membrane. This is most elegantly demonstrated by KABACK's (1969, 1970) work with osmotically active membrane vesicles derived from *E. coli* cell membranes. KABACK found that the uptake and transport of glucosides from the outside medium to the inside of these vesicles depended upon the enzymatic activity of a phosphoenolpyruvate phosphate transferase. This transferase system depended in turn upon the enzymatic phosphorylation of the glucoside in the membrane prior to its entry into the vesicle. Because diffusion could not account for the observed uptake kinetics, KABACK concluded that external sugar reached a catalytic site within the membrane whereupon it was translocated as a result of a vectorial phosphorylation.

Most interesting was KABACK's (1970) demonstration that the transport properties of the cell membrane could be functionally dissociated from the barrier properties of the membrane. At 27° C the membranes showed a steady state level of glucoside uptake and phosphorylation, whereas at 46° C rapid initial uptake was followed by leakage of the phosphorylated glucoside. Further investigations demonstrated that this leakage could be initiated instantaneously by allowing the membranes to accumulate the phosphorylated glucoside at 27° C and then raising the temperature to above 40° C. Even though the membranes became leaky above 40° C, their enzymatic transport properties increased exponentially with temperature up to 46° C. Variation in osmolarity of the surrounding medium also dissociated the transport and barrier properties of the membrane. Under optimal osmotic conditions the membrane matrix retained the small molecules which had been transported, whereas at low osmolarities the vesicles swelled and stretched, allowing leakage of the transported substrates. In no case did the osmotic treatments, which clearly modified the permeability of the membrane, alter its ability to carry out transport, *i.e.*, vectorial translocation. Other experiments using non-ionic detergents and phospholipases also showed that the barrier functions of the membrane could be altered without modifying its transport properties. The effectiveness of detergents and lipases in destroying the barrier functions and the reversibility of temperature effects on barrier functions can be explained by postulating that the barrier properties are properties of the membrane matrix. The lipids of the matrix would undergo abrupt and reversible temperature dependent phase changes only if they were disposed so as to allow for extensive lipid-lipid associations. This would be the case if the lipids were in a bilayer. On the other hand, the enzymatic transport properties of the membrane were relatively insensitive to many of the treatments which affected membrane permeability. This can be explained if the vectorial transport properties belong to distinct proteins which are differentiated from but perhaps buried within the bilayer lipid matrix of the membrane. Although some specific lipid-protein interaction seems to be required for the enzymatic activity of these transport proteins, this interaction appears to involve only a small portion of the total membrane lipid.

Are there more direct structural grounds for postulating differentiated regions within the plane of cellular membranes? Although various spectroscopic and diffraction techniques provide a refined image of the general molecular organization within the membrane, it is clear that these averaging techniques have not been used to probe the structure in limited, specific parts of the membrane. It is here that electron microscopy, and particularly the relatively new freeze-etch method, may help us.

B. Freeze-Etching

Whereas thin sectioning methods provide information about the structure in a section of a membrane, the freeze-etch technique allows us to examine structure within the plane of the membrane. Unlike most electron microscope preparatory techniques, freeze-etching does not require the use of chemical fixatives and dehydrating agents. Tissues are rapidly frozen and then fractured under vacuum (Fig. 8). After fracture a small amount of ice may be sublimed (etched) from the fractured face, and the etched surface is then shadowed and replicated. Examination of the replica in the electron microscope reveals extensive face views of membranes. The

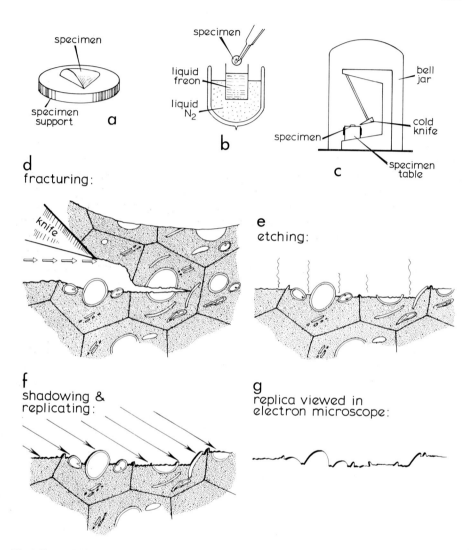

Fig. 8. Freeze-etching. The fresh specimen is a) placed on a copper disk, b) rapidly frozen in liquid Freon 22, and c) placed in the precooled freeze-etching vacuum chamber. d) The frozen specimen is fractured with a cooled microtome knife, and, in some cases, e) the freshly fractured surface is etched. f) The surface is shadowed and replicated with platinum and carbon. After dissolving away the specimen, the remaining replica g) is examined in the electron microscope.

freeze-etch technique is particularly valuable as a counterpart to biochemical fractionation and separation of membrane components. Many of these biochemical approaches involve the weakening of hydrophobic associations followed by purification of the resulting components. Freeze-etching also depends upon a weakening of hydrophobic interactions, but it is followed by direct examination of the membrane components in the electron microscope.

Whereas biochemical approaches usually depend upon detergents to weaken hydrophobic bonds, freeze-etching utilizes the freezing process to accomplish the same end. Freezing weakens hydrophobic bonds because, unlike many types of polar bonding, hydrophobic bonding involves large entropy changes attributable to structural restrictions on the water surrounding non-polar groups (FRANK and EVANS 1945, KAUZMANN 1959). Hydrophobic bonds are weaker in frozen specimens (DEAMER and BRANTON 1967) where there is no question either of surrounding newly exposed polar groups with water or of structural reorganization in the frozen aqueous phase.

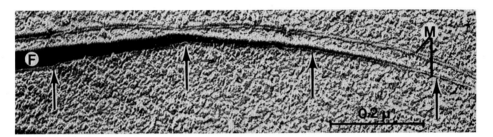

Fig. 9. Freeze-etched onion root tip cell showing portions of two endoplasmic reticulum membranes (*M*). The fractures are tangent to the membrane on the left and almost perpendicular to the membrane surface on the right. The small ridge (arrows) at the base of an exposed membrane face on the left (*F*) is continuous with one of the ridges which forms the typical freeze-etch image of a single cross fractured membrane on the right. ×120,000. From BRANTON (1966).

Fracture of a frozen, hydrophobically bonded structure has been demonstrated in experiments with transferred stearate layers (DEAMER and BRANTON 1967). Bilayers and multilayers of carbon-14-labeled stearate were frozen in contact with an aqueous phase and then fractured. The distribution of radioactivity on both sides of the fracture showed that the stearate layers were cleaved apart predominantly in the plane of their hydrocarbon tails. Although a stearate bilayer is obviously not a biological membrane, it is similar in that it includes an extensive hydrophobic region, consisting of hydrocarbon chains, and a hydrophilic region, composed of carboxylate groups. Hence, the stearate bilayer may be considered a model structure which demonstrates the important role of hydrophobic bonding in determining the location of the fracture plane.

A consistent feature of freeze-etched biological membranes is a small ridge at the base of their exposed faces (BRANTON 1966, 1967). This small ridge is, in fact, continuous with and identical to one of the two ridges which form the characteristic freeze-etch image of cross fractured membranes (Fig. 9). This type of fracture has been observed in freeze-etched preparations of a variety of plant and animal membranes (BRANTON and PARK 1967, MEYER and WINKELMAN 1969) and has been interpreted to mean that fracture splits frozen membranes, exposing either one or the other of their two inner hydrophobic faces. This interpretation implies that the fracture process of freeze-etching exposes neither the true membrane surface nor the surface of any contiguous materials. These implications have been verified by Pinto da Silva and Branton (1970) who used membrane surface markers and supplemented the usual fracture process with extensive etching to lower the ice table around the unfractured portions of the membranes.

The rational for these experiments is illustrated in Fig. 10. It was reasoned that if the fracture passed through the hydrophobic matrix of a membrane, neither the membrane surface nor any surface attached markers would be exposed. However, if fracture were followed by etching to sublime away the surrounding ice, the surface with any attached surface labels would be exposed in addition to the fracture face. A ridge re-

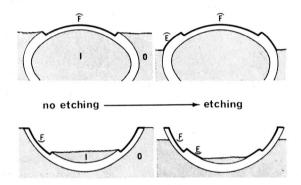

Fig. 10. Rationale of fracturing and etching, assuming membrane splitting. Without etching, fracture exposes convex ($\widehat{F}$) and concave (F) fracture faces. As the ice inside (I) and outside (O) the cell is lowered by etching, convex ($\widehat{E}$) and concave (E) etch faces are exposed. From Pinto da Silva and Branton (1970).

presenting a portion of the membrane thickness would form the border between the fracture face and the etch face.

With this reasoning in mind, etching experiments were performed on red blood cell ghosts to which ferritin label had been conjugated (Pinto da Silva and Branton 1970). When these cells were freeze-fractured and etched, ferritin was clearly visible on the etch faces but not on the fracture faces (Fig. 11). The fracture faces were identical to those of control cells without conjugated ferritin, indicating that only the etch faces and not the fracture faces showed the morphological features associated with the membrane surface. Thus the structural detail seen on fracture faces cannot be that of the membrane surface, and fracture must expose structure within the matrix of the membrane. Similar experiments and conclusions have been derived from actin-labelled erythrocyte ghosts (Tillack and Marchesi 1970) and from isolated sea urchin nuclei whose surfaces are naturally "labelled" with ribosomes (Wartiovaara and Branton 1970).

Two further observations indicate that the membrane faces seen in freeze-etching are not membrane surfaces:

1. Lipid extraction of tissue completely eliminates the planes of weakness which give rise to the characteristic fracture faces of membranes (Branton and Park 1967). This is true even though lipid extraction does not destroy the general shape and morphology of the membranes (Fleischer

et al. 1967, PARK and BRANTON 1966). Since the presence of lipid appears to be necessary to establish planes of weakness in the frozen specimen, cleavage must occur within the non-polar interior of membranes.

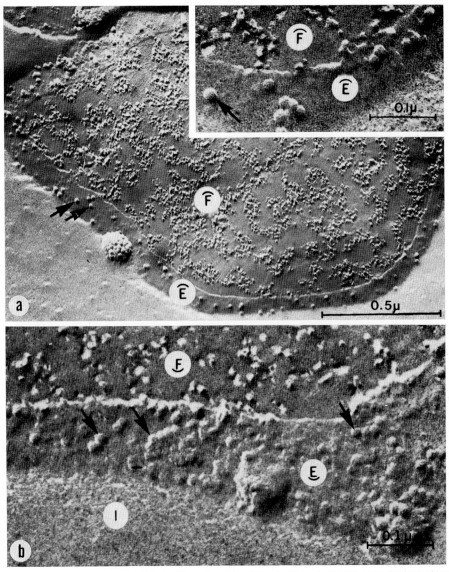

Fig. 11. Membrane faces of an etched, ferritin-conjugated red blood cell ghost. Ferritin molecules (arrows) are associated with the etched faces ($\widehat{E}$ and $\underset{\smile}{E}$) but are not visible on the fracture faces ($\widehat{F}$ and $\underset{\smile}{F}$). Symbols as in Fig. 10. Fig. 11 a, Convex fracture. ×63,000; inset, ×180,000; Fig. 11 b, Concave fracture. ×180,000. From PINTO DA SILVA and BRANTON (1970).

2. Osmium fixatives, which react with unsaturated but not saturated fatty acids (STOECKENIUS and MAHR 1965), destroy the natural fracture planes of membranes containing unsaturated fatty acids so that few face views

of the fixed membranes are visible in freeze-etch preparations (Fig. 12 d) (JAMES and BRANTON 1971). Osmium fixation of membranes containing

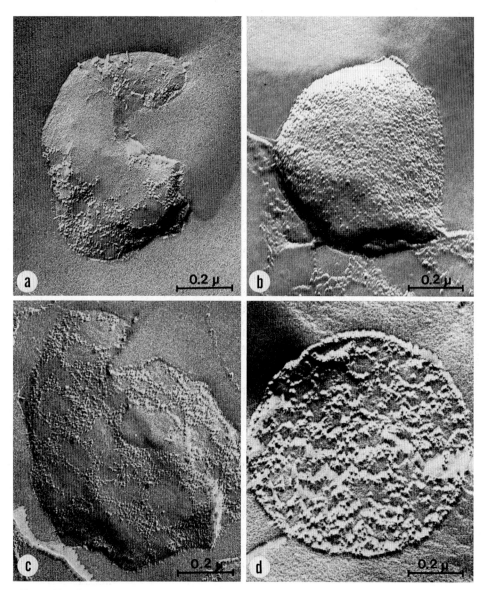

Fig. 12. The effect of OsO₄ fixation on the fracture process in *Mycoplasma laidlawii* membranes enriched with saturated and unsaturated fatty acids. Top, unfixed cell membranes containing *a*) less than 0.25 double bond per fatty acid and *b*) over 0.60 double bond per fatty acid. The stearate supplemented cells are less regular in shape and bear fewer particles than the more coccoid linoleate supplemented cells. Bottom, OsO₄ fixed (*c*) stearate and (*d*) oleate supplemented cells. Fixation does not affect the fracture plane of stearate supplemented cells, but it causes an increased number of cross fractures in linoleate supplemented cells. All ×72,000. From TOURTELLOTTE et al. (1970) and JAMES and BRANTON (1971).

primarily saturated fatty acids does not destroy the fracture planes and yields the same membrane faces as the unfixed controls (Fig. 12 c). Therefore, the freeze-etch membrane faces must be the inner, hydrophobic regions of the membrane which are susceptible to osmium fixation when they contain unsaturated fatty acids. On the other hand, aldehyde fixatives are known to effect cross-linkage of hydrophilic portions of proteins (SABATINI

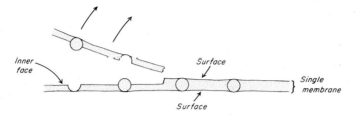

Fig. 13. Interpretation of the fracture process. The inner membrane faces, not the membrane surfaces, are exposed by the fracture process of freeze-etching.

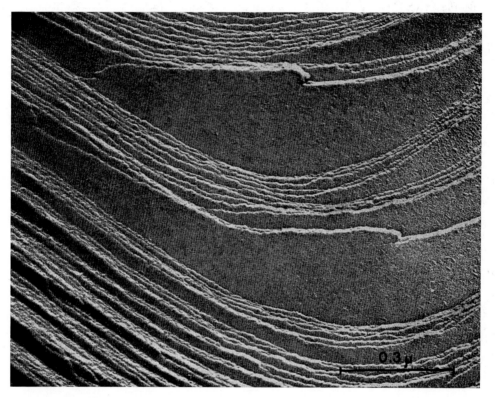

Fig. 14. Five-day-old rat sciatic nerve myelin. Exposed fracture faces are smooth. ×100,000. From BRANTON (1967).

et al. 1963, HABEEB and HIRAMOTO 1968). Numerous aldehyde-fixed structures have been freeze-etched (JOST 1965, BULLIVANT and AMES 1966, PARK and BRANTON 1966) and found to yield the same membrane faces as the unfixed controls. This again suggests that the freeze-etch membrane faces are the

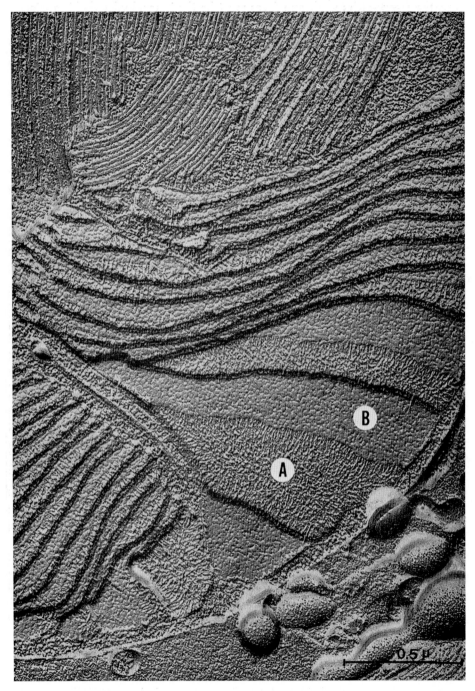

Fig. 15. Oblique fracture through a guinea pig retinal rod outer segment showing parts of several disc membrane fracture faces. One of the faces (arbitrarily labeled "A") has the appearance of shallow, irregularly shaped pits surrounded by steep, interconnecting ridges. The other face (arbitrarily labeled "B") has the appearance of worn cobblestone pavement. ×63,000. From BRANTON and CLARK (1968).

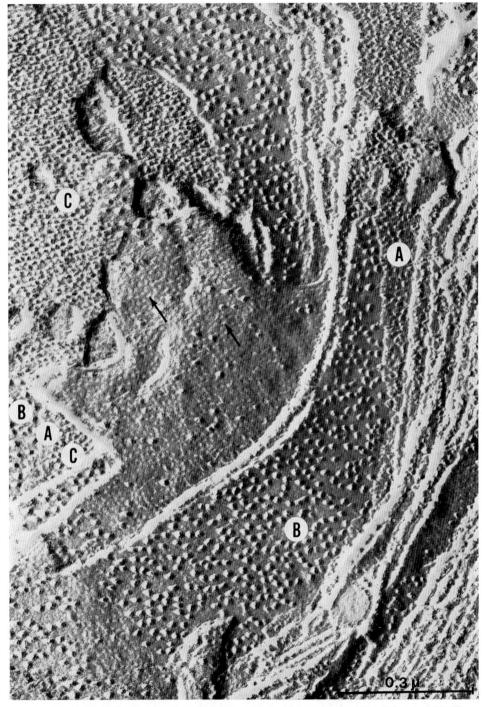

Fig. 16. Face views of chloroplast lamellae. Three different kinds of faces can be seen. They have been arbitrarily labeled "A", "B", and "C". Face A is a rough surface with no discrete particles. Face B appears as a relatively smooth background upon which are distributed particles 160 Å to 200 Å in diameter. Arrows point to depressions from which these particles have been pulled away during the fracture process of freeze-etching. Face C appears as an array of many discrete particles, mostly 100 Å to 130 Å in diameter but with occasional particles up to 200 Å in diameter. Isolated spinach chloroplasts, resuspended in 10% glycerol. ×120,000. From BRANTON and PARK (1967).

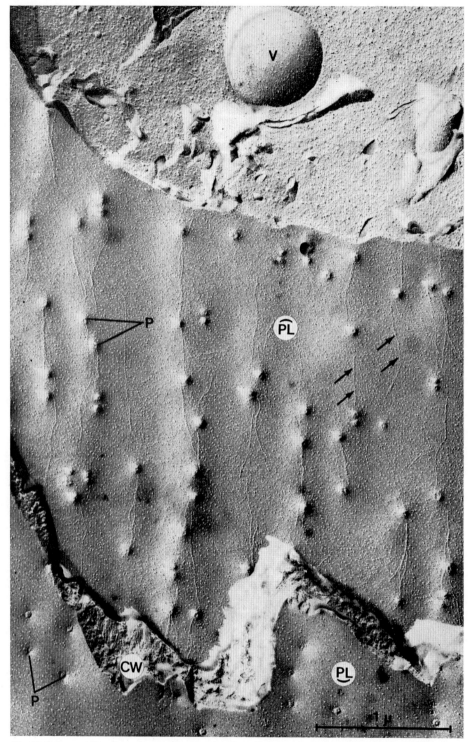

Fig. 17. Onion root tip cell membrane. The cell cytoplasm with a vacuole (V) is visible at the top of the photograph, the convex cell membrane (plasmalemma, $\overparen{PL}$) at the center. This plasmalemma has been exposed by a fracture which has split away a large portion of the cell wall (CW) whose edge is visible toward the bottom of the photograph. The concave fracture face of the neighboring cell membrane $(\underparen{PL})$ is at the bottom. Broken off plasmadesmata (P) are visible on both exposed plasmalemmae. The membrane particles are randomly distributed throughout the membrane and also occur in files (arrows). ×36,000.

inner, hydrophobic areas of the membrane and not the outer, hydrophilic, glutaraldehyde-susceptible surfaces.

The ability to split membranes has been particularly helpful in determining the existence, the extent and the distribution of any structurally differentiated regions within the hydrophobic matrix. In most membrane systems differentiated regions do exist and they appear in the freeze-etch replicas as small particles in the otherwise smooth continuum of fractured membrane matrix (Fig. 13).

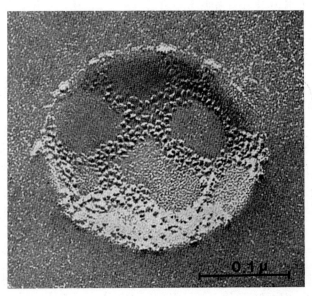

Fig. 18. Microsomal vesicle from lobster tail muscle. The vesicle has cleaved to produce a concave face and has numerous 80—90 Å particles embedded in the fracture face. (In these preparations convex fracture faces are typically smooth.) Such vesicles have calcium transport activity and are therefore believed to be derived from membranes of the sarcoplasmic reticulum. ×234,000.

A large number of plant and animal membranes has been examined by the freeze-etch technique (for reviews see KOEHLER 1968, MOOR 1966). Of these only myelin (Fig. 14) appears to be smooth and devoid of particles or discontinuities. Retinal rod discs (Fig. 15) and chloroplast lamellae (Fig. 16) appear unusually complex and rich in these discontinuities. Other membrane types fall between these two extremes, and, in fact, the particle number and distribution are characteristic of particular membrane types. Some of these membranes are illustrated in Figs. 17–20, and data regarding particle density in a number of different membrane systems are assembled in Table 3. Although the value of this compilation is limited by the fact that the data were taken from cells under different growth conditions and at different stages of the cell cycle, some of the more striking trends are noteworthy.

In the first place, in every membrane system for which the data is available there is a marked and consistent asymmetry in the distribution

Table 3. *Particle Density on Fractured Membrane Faces*

Type of Membrane	Number of particles per μ^2 of membrane face		% Membrane area covered by particles[a]	Reference[b]
	Densely populated face	Thinly populated face		
Lecithin myelin forms	0	0	0	Staehelin 1968*
Myelin sheath	0	0	0	Branton 1967*
Mitochondrial inner membrane	2700	—	—	Moor 1964; Moor et al. 1964
Endoplasmic reticulum (root tip)	1700	380	12	Branton, unpublished
Nuclear membrane (root tip)	1790	420	12	Branton and Moor 1964*
Vacuole membrane (root tip)	3300	2480	32	Branton 1966; Branton and Moor 1964; Mattle and Moor 1968*
Muscle microsomes	4300	0	35	Deamer, D. W., unpublished
Chloroplast lamellae	3860	1800	80	Branton and Park 1967; Mühlethaler et al. 1965*
Plasma membranes				
Mycoplasma laidlawii B				
Oleate supplemented cells	3600	900	16	Tourtellotte et al. 1970
Stearate supplemented cells	1600	400	7	Tourtellotte et al. 1970
	2800	700	12	James and Branton 1971*
Onion root tip	2030	550	15	Branton and Moor 1964
				Northcoat and Lewis 1968*
Human erythrocyte	2800	1400	23	Weinstein and Bulivant 1967*
				Mattle et al. 1967;
Bakers' yeast	2600	—	63	Moor and Mühlethaler 1963*

[a] Area covered by particles calculated from total number of particles and measured diameter of individual particles.
[b] Where reference is followed by an asterisk, particle counts made on published photographs have been supplemented, for greater accuracy, by counts on similar unpublished photographs in one of the author's (D.B.) files.

of particles between the two fracture faces, indicating that the particles are more strongly bound to one half of the membrane than to the other (Fig. 21). As the two halves are cleaved apart, the particles on one face leave some small, hard to discern holes or depressions on the previously

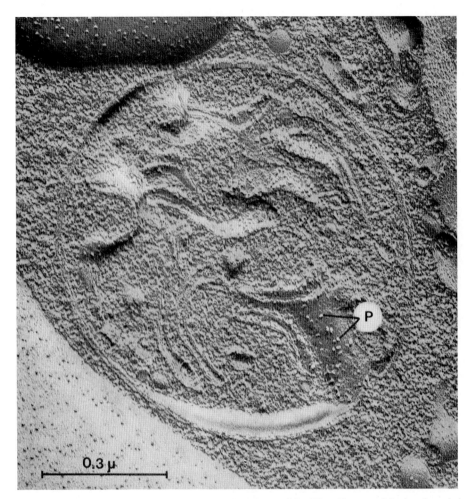

Fig. 19. Onion root tip mitochondrion. There are relatively few particles (*P*) on the exposed fracture faces, but the number of these particles varies greatly in different mitochondria. ×105,000.

apposed face (see Fig. 16). The pattern of asymmetric particle distribution between the two membrane faces reflects well-known membrane relationships, *e.g.*, continuity between nuclear membrane and endoplasmic reticulum (ROBERTSON 1959) and between endoplasmic reticulum and tonoplast (MATILE and MOOR 1968). It is particularly interesting to note that the relative particle distribution between the two membrane faces in mitochondrial cristae and chloroplast lamellae is consistent with the view that the inner membranes of these organelles are analogous to autonomous cell membranes.

It is also noteworthy that the number of particles in the membrane matrix is greatest in "active" membranes such as the chloroplast lamellae,

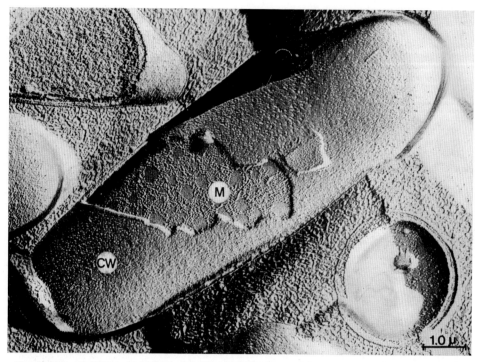

Fig. 20. Cell wall (*CW*) and membrane face (*M*) of *E. coli*. The membrane particles are rather closely packed around circular, bare patches which are devoid or particles. From FIIL and BRANTON (1969). × 11,000.

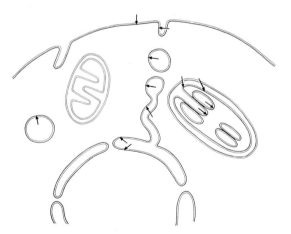

Fig. 21. Membrane asymmetry. Using the data compiled in Table 3, the fracture face which always has the largest number of particles is shown by arrows. From BRANTON (1969).

which perform the light reactions of photosynthesis (PARK and PON 1961, TREBST *et al.* 1958), and least in "inactive" membranes such as the myelin layers, which function primarily as metabolically inert insulators around the axon (DAVISON and DOBBING 1960, O'BRIEN 1965). Such observations suggest that the functional complexity of a membrane may be directly related to number of particles, a suggestion which immediately begs the question, "What do the smooth areas and particles represent in terms of specific chemical or functional membrane components?"

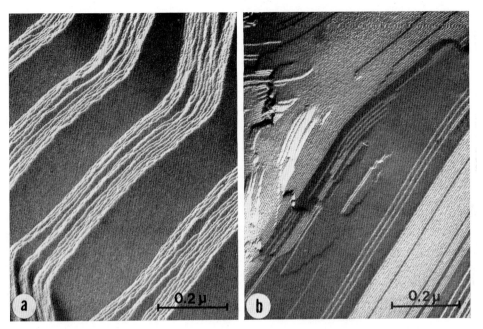

Fig. 22. Fractures in a) lamellar phase (L-α-dipalmitoyl) and b) hexagonal II phase (calcium cardiolipin) lipids. ×90,000. From DEAMER et al. (1970).

One approach to this question has been the study of model systems composed of individual membrane components. Considering that fracture between the methyl-ends of membrane phospholipids may account for the extended smooth regions seen in membranes (DEAMER and BRANTON 1967), DEAMER et al. (1970) examined a range of bulk phase fatty acids and phospholipids. In every case where the existence of a lamellar phase structure was indicated by an independent observation such as X-ray diffraction, the freeze-etch results showed extensive smooth faces similar to the smooth regions seen in biological membranes (Fig. 22 a). Examinations of phospholipids with saturated and unsaturated fatty acids and with differing polar groups showed that neither the degree of saturation nor the admixture of other components such as cholesterol altered the freeze-etch image.

Before concluding that the extended smooth sheets seen in freeze-etched biological specimens were diagnostic of lamellar phase lipids, DEAMER et al. (1970) showed that other phases produced a different appearance when freeze-etched. Several lipid systems occur in hexagonal phases as defined by X-ray diffraction (Luzzati 1968). In these hexagonal phases the lipid

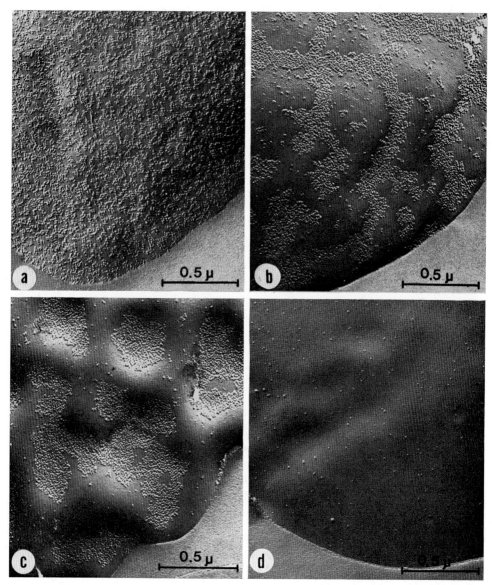

Fig. 23. Particle aggregation and removal during pronase digestion of red blood cell ghosts. (*a*) Control cell incubated in buffer only; *b*), *c*) and *d*) digested cells incubated in pronase so as to remove *b*) 30%, *c*) 45% or *d*) 70% of the original protein. All ×40,000. From Engstrom (1970).

molecules are arranged in rods of indefinite length; the packing of the rods is hexagonal. The freeze-etch images of all the hexagonal phase lipids examined were distinctly different from those of lamellar phase lipids and were composed of indefinitely long parallel lines, giving the fracture face a ribbed appearance (Fig. 22 b). Furthermore, each specimen contained at least two fracture planes at approximately 120° to each other.

These observations show that consistent and recognizable features distinguish the freeze-etch images of different lipid phases. Because the bulk

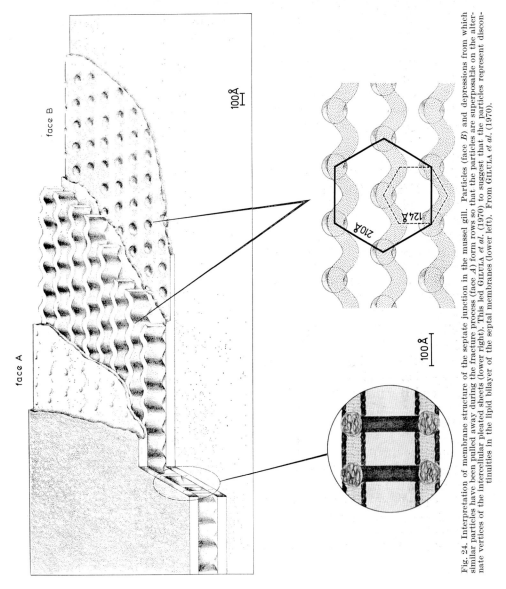

Fig. 24. Interpretation of membrane structure of the septate junction in the mussel gill. Particles (face B) and depressions from which similar particles have been pulled away during the fracture process (face A) form rows so that the particles are superposable on the alternate vertices of the intercellular pleated sheets (lower right). This led GILULA et al. (1970) to suggest that the particles represent discontinuities in the lipid bilayer of the septal membranes (lower left). From GILULA et al. (1970).

phase lipids, even when hydrated, did not contain any discontinuities which might contribute to the particulate appearance of fractured biological membranes, DEAMER et al. (1970) suggested that the particles in biological membranes must be due to proteins or localized and specific protein-lipid interactions. This suggestion has been investigated by using freeze-etching to follow the course of proteolytic digestion in red blood cell ghosts (ENGSTROM 1970). If the particles represent proteins within the membrane matrix, they should disappear when protein is removed from the ghosts prior to freeze-etching. Red blood cell ghosts were incubated in the proteolytic enzyme pronase for varying time periods. The course of protein hydrolysis was monitored by simultaneously measuring protein loss and particle loss. Loss of membrane protein initially caused extensive particle aggregation with little decrease in particle number (Figs. 23 a, b, and c). This initial resistance of the particles to pronase digestion supports the view that the particles are buried within a hydrophobic region of the membrane matrix which would retard the diffusion of pronase. Most of the particles were lost after prolonged pronase treatment (Fig. 23 d), at which point the membranes began to vesiculate and disintegrate. The eventual removal of particles from extensively digested ghosts supports the hypothesis that the particles contain protein.

Attempts to identify the particles by chemical extraction have also been made by MARCHESI (1970) and TILLACK et al. (1970). They found that lithium diiodosalicylate removed a major glycoprotein fraction from erythrocyte membranes. Extraction of erythrocytes with lithium diiodosalicylate caused the particles to disappear.

More direct chemical characterization of the particles has come from attempts to label the A antigen sites of the ABO antigen system on erythrocyte membranes. These experiments (PINTO DA SILVA et al. 1970, 1971) utilized freeze-etching techniques to label human erythrocyte A antigen sites with ferritin conjugated to anti-A antibodies. When the ferritin labelled antibodies were bound to the erythrocyte surface, the ferritin molecules, and hence the anti-A antibodies, were always localized on those areas of the membrane surface which overlay the membrane particles. PINTO DA SILVA et al. concluded that since immunochemical evidence indicates the A and B antigens are glycoprotein and glycolipid, the correspondence between the antigen sites and the membrane particles implies that the particles contain glycoprotein or glycolipid moieties. PINTO DA SILVA et al. also noted that although most of the membrane particles were preferentially associated with the inner half of the membrane (see Fig. 23); they must traverse the entire membrane and protrude to the outer surface where they carry the ABO antigen determinant.

The membrane particles may also serve other functions. Indeed, there is evidence that in certain cases they may be an important part of the low resistance pathway that accounts for electrical coupling between animal cells. This evidence is based on work by GILULA et al. (1970) who investigated the septate junction in mussel gills. This junction is a 2–3 μ wide, belt-like structure which occurs between gill epithelial cells. The cell membranes of the junction are separated by a 150 Å intercellular space which is regularly traversed by septa that join the membranes of two adjacent cells. Using conventional electron microscopy, GILULA et al. showed that the septa are pleated sheets which are structurally continuous with the membrane surfaces (Fig. 24). Within the hydrophobic matrix of these membranes freeze-etching exposed geometrically arranged rows of 85 Å particles. Most important, GILULA et al. showed that these membrane particles were superposable on the alternate vertices of the intercellular septal sheets (Fig. 24, lower right). Thus, the septal sheets together with the membrane particles appear to provide a channel that extends from the cytoplasm of one cell through the septate junction to the cytoplasm of the adjacent cell. These studies are among the first which make it possible to ascribe a discrete function, in this case cellular coupling, to the membrane particles exposed by freeze-etching.

IV. Conclusions

The evidence we have reviewed suggests that the time averaged structure of biological membranes is that of an asymmetric, protein-covered lipid bilayer within which are intercalated specific, functionally differentiated, protein rich sites. As we have seen, the ideas behind this generalized PLP structure are not new but have evolved gradually, first from attempts to account for membrane permeability and composition and more recently from a variety of spectroscopic and electron microscopic probes. This PLP model provides a structural basis for suggestions that the unspecialized permeability properties of a membrane will be attributes of a lipid bilayer. Furthermore, the model ascribes specialized permeability, transport and functional properties of a membrane to a limited number of discrete sites which may represent only a fraction of the total membrane volume. This explains why averaging techniques which measure overall structure do not account for many aspects of membrane activity and specificity. Finally, the model provides for a hydrophobic environment to accommodate enzymes which cannot function in an aqueous milieu. Enzymes coupling oxidation-reduction and phosphorylation are examples of the kind of proteins which, because they mediate hydration-dehydration reactions, might be intercalated within the non-aqueous environment of the lipid bilayer.

The freeze-etch studies we have described help to focus our understanding of the PLP membrane in terms of a concrete picture. The observation that the fractures used in this technique tend to be propagated along all biological membranes and in fact appear to split most biological membranes is in itself persuasive evidence for the planar continuity and bilayered arrangement of these membranes. On the other hand, the fact that particles are exposed by fracturing, particularly in metabolically active membranes, suggests that a simple lipid bilayer arrangement does not pervade the entire membrane. Localized protein intercalations or some other differentiation of the lipid bilayer must be invoked to account for these particulate structures. Keeping in mind that in some cases virtually none of the membrane area is occupied by these particles whereas in others up to 80 per cent of the area is particulate, any generalized scheme must be sufficiently flexible to allow for this type of structural variation between different membranes.

Indeed, our generalized PLP model includes some of the major features underlying LPL and particulate membrane models. For instance, some of the protein is assumed to be intercalated within the hydrophobic interior of the membrane, and this protein, if it could be isolated and dispersed with a suitable amount of lipid, might form lipid-protein aggregates which could be described as membrane subunits. But, as our review of the available evidence shows, only a limited proportion of any membrane's protein appears to be buried within its hydrophobic core, and convincing evidence for extensive hydrophobic interaction between lipid and protein has not been found.

Although there is a considerable body of evidence supporting a generalized PLP model, there is far less evidence regarding the details of lipid-protein interaction that exists within this framework. It is therefore not at all surprising that even the most recent membrane models (SJOSTRAND and BARAJAS 1970, VANDERKOOI and GREEN 1970, SINGER 1971), all of which contain some type of lipid bilayer, imply completely divergent views about how the two major membrane components—lipids and proteins—are held together. In large part, the divergence of these views may be attributed to the tendency of each researcher to treat those particular membrane proteins and lipids with which his studies have been concerned as though they were representative of all membrane proteins and lipids. In fact, at this stage of our knowledge, it is probably misleading to speak of "membrane proteins" or "membrane lipids" as though each was contained in a well mixed, bulk phase. It is true that we can now provide strong arguments regarding the arrangement of the bulk of the membrane components, but our preoccupation with models showing an average membrane structure must give way to more exact and detailed visions. Clearly, future progress in understanding membrane structure will come from studies directed toward individual proteins and lipids, their precise modes of interaction, and their specialized relations to the topology and dimensions of individual cellular membranes.

Acknowledgements

We are most grateful to Eleanor Crump for her dedicated help in assemblying this manuscript. Thanks to her skill, problems of intercontinental coordination between sabbatical authors were easily solved. This manuscript was completed while D. B. was a Fellow of the John Simon Guggenheim Memorial Foundation. Original research reported in this review was supported by A. E. C. contract AT (04–3)–34, P. A. 142 (D. B.) and NSF grants GB 8600 and 23186 (D. W. D.).

Bibliography

ARNOLD, W., 1933: The effect of ultraviolet light on photosynthesis. J. gen. Physiol. **17**, 135.

ARNTZEN, C. J., R. A. DILLEY, and F. L. CRANE, 1969: A comparison of chloroplast membrane surfaces visualized by freeze-etch and negative staining techniques; and the ultrastructural characterization of membrane fractions obtained from digitonin-treated spinach chloroplasts. J. Cell Biol. **43**, 16.

BADER, H., R. POST, and G. BOND, 1968: Comparison of sources of a phosphorylated intermediate in transport ATPase. Biochim. biophys. Acta **150**, 41.

BAR, R. S., D. W. DEAMER, and D. G. CORNWELL, 1966: Surface area of human erythrocyte lipids: reinvestigation of experiments on plasma membrane. Science **153**, 1010.
BELLAMY, L. J., 1958: Infrared Spectra of Complex Molecules. London: Methuen.
BENSON, A. A., 1964: Plant membrane lipids. Ann. Rev. Plant Physiol. **15**, 1.
— and S. J. SINGER, 1965: The lipoprotein model for membrane structure. Amer. Chem. Soc. Abstracts of Papers **150**, 8 c.
BRANTON, D., 1966: Fracture faces of frozen membranes. Proc. nat. Acad. Sci. U.S. **55**, 1048.
— 1967: Fracture faces of frozen myelin. Exp. Cell Res. **45**, 703.
— 1968: Structure of the photosynthetic apparatus. Photophysiology **3**, 197.
— 1969: Membrane structure. Ann. Rev. Plant Physiol. **20**, 209.
— and A. W. CLARK, 1968: Fracture faces in frozen outer segments from the guinea pig retina. Z. Zellforsch. **91**, 586.
— and H. MOOR, 1964: Fine structure in freeze-etched *Allium cepa* L. root tips. J. Ultrastruct. Res. **11**, 401.
— and R. B. PARK, 1967: Subunits in chloroplast lamellae. J. Ultrastruct. Res. **19**, 283.
BULLIVANT, S., and A. AMES, 1966: A simple freeze-fracture replication method for electron microscopy. J. Cell Biol. **29**, 435.

CASPAR, D. L. D., and D. A. KIRSCHENER, 1971: Myelin membrane structure at 10 Å resolution. Nature **231**, 46.
CHAPMAN, D., D. J. FLUCK, S. A. PENKETT, and G. G. SHIPLEY, 1968 a: Physical studies of phospholipids. X. The effect of sonication on aqueous dispersions of egg lecithin. Biochim. biophys. Acta **163**, 255.
— V. B. KAMAT, J. DEGIER, and S. A. PENKETT, 1968 b: Nuclear magnetic resonance studies of erythrocyte membranes. J. molec. Biol. **31**, 101.
— — and R. J. LEVENE, 1968 c: Infrared spectra and the chain organization of erythrocyte membranes. Science **160**, 314.
— R. B. LESLIE, R. HIRZ, and A. M. SCANU, 1969: 220 MHz nuclear magnetic resonance spectra of high density serum lipoprotein. Nature **221**, 260.
— and S. A. PENKETT, 1966: Nuclear magnetic resonance spectroscopic studies of the interaction of phospholipids with cholesterol. Nature **211**, 1304.
— and D. F. H. WALLACH, 1968: Recent physical studies of phospholipids and natural membranes. In: Biological Membranes, Physical Fact and Function, 125. CHAPMAN, D., ed., New York: Academic Press.
— R. M. WILLIAMS, and B. D. LADBROOKE, 1967: Physical studies of phospholipids. VI. Thermotropic and lytotropic mesomorphism of some 1,2-diacylphosphatidylchloines (lecithins). Chem. Phys. Lipids **1**, 445.
CLIFFORD, J., and B. A. PETHICA, 1964: Properties of micellar solutions, Part 2: N. M. R. chemical shift of water protons in solutions of sodium alkyl sulphates. Trans. Faraday Soc. **60**, 1483.
COLEMAN, R., and J. B. FINEAN, 1966: Preparation and properties of isolated plasma membranes from guinea pig tissues. Biochim. biophys. Acta **125**, 197.
— J. B. FINEAN, S. KNUTTON, and A. R. LIMBRICK, 1970: A structural study of the modification of erythrocyte ghosts by phosphopholipase c. Biochim. biophys. Acta **219**, 81.
CRANE, F. L., J. W. STILES, K. S. PREZBINDOWSKI, F. J. RUZICKA, and F. A. SUN, 1968: The molecular organization of mitochondrial cristae. In: Regulatory Functions of Biological Membranes, 21. JARNEFELT, J., ed., New York: Elsevier.

DANIELLI, J. F., 1954: The present position in the field of facilitated diffusion and selective active transport. Colston Papers **7**, 1.
— and H. DAVSON, 1935: A contribution to the theory of permeability of thin films. J. Cell comp. Physiol. **5**, 495.
DAVISON, A. N., and J. DOBBING, 1960: Phospholipid metabolism in nervous tissue, 2. metabolic stability. Biochem. J. **75**, 565.
DEAMER, D. W., 1970: An alternative model for molecular organization in biological membranes. Bioenergetics **1**, 237.
— and D. BRANTON, 1967: Fracture planes in an ice-bilayer model membrane system. Science **158**, 655.
— R. LEONARD, A. TARDIEU, and D. BRANTON, 1970: Lamellar and hexagonal lipid phases visualized by freeze-etching. Biochim. biophys. Acta **219**, 47.
DE BERNARD, L., 1958: Associations moleculaire entre les lipides. Bull. Soc. Chim. Biol. **40**, 161.

Protoplasmatologia

Volume V/4

The Nuclear Structures of Protocaryotic Organisms (Bacteria and Cyanophyceae)

By Dr. **G. W. Fuhs**, Albany, N.Y.

"... This is in most respects an excellent little book. It has its faults but these are far outweighed by its virtues. The accounts of the nucleus in the introductory chapters of general textbooks are usually right up to date; in fact much too up to date, because they adopt the views and retail the information of the most recent workers in the field, and ignore everything else. Fuhs has made an impressive effort to absorb and digest the great mass of work published in this field over the past thirty years and more." *Nature*

86 figures. IV, 186 pages. 1969.
Cloth
S 573,—, DM 83,—, US$ 25.90

Subscription price cloth
S 458,—, DM 66,40, US$ 20.70

Volume VI/A

The Chromosome Complement

By Dr. **B. John**, Birmingham, and Dr. **K. R. Lewis**, Oxford

87 figures. IV, 206 pages. 1968.
Soft cover
S 607,—, DM 88,—, US$ 27.40

Subscription price soft cover
S 486,—, DM 70,40, US$ 21.90

Volume VI/B

The Chromosome Cycle

By Dr. **B. John**, Birmingham, and Dr. **K. R. Lewis**, Oxford

45 figures. IV, 125 pages. 1969.
Cloth
S 497,—, DM 72,—, US$ 22.40

Subscription price cloth
S 398,—, DM 57,60, US$ 18.00

Volume VI/F/1

The Meiotic System

By Dr. **B. John**, Birmingham, and Dr. **K. R. Lewis**, Oxford

195 figures. IV, 335 pages. 1965.
Soft cover
S 997,—, DM 144,50, US$ 45.00

Subscription price soft cover
S 798,—, DM 115,60, US$ 36.00

SPRINGER-VERLAG WIEN · NEW YORK

THE LIBRARY